Reviews of Physiology Biochemistry and Pharmacology 129

Springer-Verlag Berlin Heidelberg GmbH

Reviews of
129 Physiology Biochemistry and Pharmacology

Editors
M.P. Blaustein, Baltimore   H. Grunicke, Innsbruck
D. Pette, Konstanz   G. Schultz, Berlin
M. Schweiger, Berlin

Honorary Editor:
E. Habermann, Gießen

With 5 Figures and 8 Tables

Springer

ISSN 0303-4240

ISBN 978-3-662-30999-5   ISBN 978-3-540-68539-5 (eBook)
DOI 10.1007/978-3-540-68539-5

Library of Congress-Catalog-Card Number 74-3674

© Springer -Verlag Berlin Heidelberg 1996
Originally published by Springer-Verlag Berlin Heidelberg New York in 1996
Softcover reprint of the hardcover 1st edition 1996

Production: PRO EDIT GmbH, D-69126 Heidelberg
SPIN: 10482848   27/3136-5 4 3 2 1 0 – Printed on acid-free paper

Contents

Indexed in Current Contents

Genetic Approaches to Cancer Immunotherapy

Thomas Blankenstein[1], Sophie Cayeux[2], and Zhihai Qin[1]

1. Max-Delbrück-Centrum für Molekulare Medizin, Robert-Rössle-Straße 10, 13122 Berlin, Germany
2. Robert-Rössle Klinik, Virchow Klinikum, Humboldt University Berlin, Lindenberger Weg 80, 13125 Berlin, Germany

Contents

1
Introduction

Since the development of methods for gene transfer in mammalian cells about 15 years ago and in particular after the introduction of those with high efficiency such as viral systems the use of genetically manipulated somatic cells for therapy has continuously been such an attractive topic that it has been frequently reviewed (Anderson 1984; Kantoff et al. 1988; Friedmann 1989; Verma 1990; Miller 1992; Anderson 1992; Mulligan 1993; Counoyer and Caskey 1993). Originally, the main targets for gene therapy were inherited monogenetic recessive disorders. Examples are diseases with defects in genes encoding adenosine deaminase (severe combined immunodeficiency), β-globin (thalassemia), factor VIII, IX (hemophilia), cystic fibrosis transmembrane regulator (cystic fibrosis), glucocerebrosidase (Gaucher's disease), β-glucuronidase (mucopolysaccharidosis), low-density lipoprotein (LDL)-receptor (hypercholesterolemia) or dystrophin (Duchenne muscular dystrophy). In recent years, however, the major focus of gene therapy approaches has turned to acquired diseases, mainly cancer. Several reasons may account for that: (a) cancer patients are numerous, but patients with most genetic defects are rare, (b) it is still difficult to obtain a regulated and prolonged expression of transgenes in vivo, (c) with few exceptions such as the hematopoietic system primitive stem cells able to long term reconstitute the respective organ are not yet well characterized, and (d) cost/benefit calculations (side effects versus therapeutic benefit) seem to favor cancer for gene therapy over genetic diseases.

The latter point means that potential side effects or toxicity of gene therapy modalities are expected to be relatively low in cancer patients in comparison to traditional radiation and chemotherapy and, moreover, the traditional therapy does not provide long term cure of cancer patients in most cases. Thus, physicians are often left in the hopeless situation of being unable to offer any further therapy to their patients. However, neither is gene therapy of cancer thought to replace but to support current forms of therapy nor can a cost/benefit factor accurately be estimated at the present time because it is too early to draw any conclusions regarding therapeutic benefit. In any case, it appears that of the more than 100 clinical gene therapy protocols the vast majority is related to cancer.

Several strategies for cancer gene therapy can be distinguished: (a) gene correction, (b) 'suicide' gene strategy, (c) gene marking in a broader sense and drug resistance gene transfer and (d) immunogene therapy. Gene correction is based on the knowledge that in certain tumors malignant transformation is causally related to gene loss or expression of mutant tumor suppressor genes, e.g., p53, or, alternatively, to activated mutant protooncogenes, e.g., ras genes. Introduction of a wildtype p53 gene or inhibition of oncogene expression by antisense molecules can revert the malignant phenotype of tumor cells and eventually kill the tumor cells, presumably by apoptosis (Liebermann et al. 1995). The effective gene correction requires each individual tumor cell to be transduced in vivo, but this is difficult to achieve by current techniques. The suicide gene strategy, in principle, is a localized chemotherapy. Suicide genes under investigation are the herpes simplex virus thymidine kinase (HSV-TK) and the bacterial cytosine deaminase genes. Enzymes encoded by these genes are able to convert appropriate prodrugs into toxic compounds which selectively kill those cells expressing the suicide gene. For example, cells expressing the HSV-TK gene can be eliminated in vitro and in vivo by use of the prodrug gancyclovir. Interestingly, HSVTK$^-$ cells in the neighbourhood of HSV-TK$^+$ cells are also killed by a mechanism called 'bystander' effect which is incompletely understood. By intratumoral injection of cells which produced HSV-TK gene containing retroviruses, only a small percentage of tumor cells were transduced, however, the whole tumor mass was eliminated after systemic application of gancyclovir. This strategy seems to be preferable for local inoperable tumors such as certain glioblastomas and clinical trials are ongoing (for review, see Blaese et al. 1994). The third experimental strategy is gene marking. This per se is no gene therapy approach. It firstly aims at analysing the reason of tumor relapse after autologous bone marrow transplantation, e.g., whether relapsing tumor cells derive from the transplant or from residual tumor cells which have survived chemotherapy in patients. Secondly, gene marking is useful for understanding the biology of bone marrow reconstitution in autologous or allogeneic situations. Brenner and coworkers have used a retrovirus containing neomycin as a marker gene to molecularly label bone marrow cells from acute myeloid leukemia or neuroblastoma patients. In five out of six patients reconstituted with such cells, the marker gene could be detected after relapse. This clearly demonstrated the contribution of grafted cells to tumor relapse. These results are important because they revive the question whether and by which method bone marrow purging (selective elimination of tumor cells from the transplant) is useful. In future, different purging methods can be directly compared in the same patient by the use of distinguishable marker genes (for review, see Brenner et al. 1994).

Most experimental work has been done for immunogene therapy which is the subject of this review. Tumor cells have been genetically modified with a variety of genes encoding immunostimulatory activity, most often cytokines, in order to increase the immunogenicity of tumors and induce an antitumor immune response, both, locally and systemically. Based on experimental tumor models a number of clinical trials have begun employing cytokine gene-modified tumor cells as vaccine. In the following, we will discuss the experimental findings with gene modified tumor cells, how these results relate to previous attempts of immunotherapy and what their meaning is for clinical application.

2
Development of Concepts for Cancer Immunotherapy

Because the majority of gene therapy approaches to cancer involve immunological principles, it is important to point out that cancer immunotherapy has a long history and to demonstrate how current concepts have evolved from previous attempts and consider new profound knowledge of immunoregulation. Some anecdotal examples cited by Oettgen and Old (1991) in their review on *History of Cancer Immunotherapy* illustrate a long standing interest in cancer immunotherapy (Table 1). As bizarre as some of these examples are, some of the current approaches presumably will become in future. Four general principles have constantly been followed over time: (1) employing tumor antigens as vaccines, (2) enhancing immune response with immunomodulators, (3) making use of the products of antitumor response, namely activated immune cells or (4) antibodies (Fig. 1). A general feature of these strategies is that they have evolved from very rough to more and more sophisticated approaches even though the rational behind them has not dramatically changed. All the strategies have had at least in selected situations some success in the clinic:

- *Vaccines.* Among a large number of clinical trials with tumor cell vaccines few showed therapeutical benefit. Examples are a study in colon cancer patients with autologous tumor cell/BCG as vaccine (Hoover and Hanna 1991), or a study in melanoma patients with a mixture of three allogeneic melanoma cell lines (to increase the antigenic spectrum

Table 1. The long-standing interest in cancer immunotherapy

Year	Experiment
1774	A physician in Paris injects pus into a patient with inoperable breast cancer; tumor completely disappeared
1777	Nooth, surgeon to the Duke of Kent, injects himself repeatedly with cancer tissue from a patient
1808	Alibert, physician to Louis XVIII, injects himself with a patient's cancer extract
1898	Coley's toxin is widely used in cancer patients with remarkable success
1949	Domagk, discoverer of the sulfonamides and nobel laureate, injects himself over a 10-year period repeatedly with cancer extracts

of the vaccine) injected together with the adjuvant BCG (Morton et al. 1992).

- *Immunomodulators.* The mostly cited examples are BCG instillation in superficial bladder cancer patients (Herr 1991), interferon (IFN)-α treatment of hairy cell leukemia (Moormeier and Golomb 1991) and interleukin (IL)-2 treatment of melanoma and renal cell carcinoma (Lotze and Rosenberg 1991).
- *Immune cells.* The clearest example for therapeutic efficacy in patients may be the "graft versus tumor reaction." Following allogeneic bone marrow transplantation in cancer patients relapse occurred more frequently and rapidly when T cells were depleted from the transplant (Fefer et al. 1991). Another successful example is using in vitro expanded tumor infiltrating cells (TIL) together with IL-2 for adoptive transfer in some melanoma or renal cell carcinoma patients (Rosenberg 1991).
- *Antibodies.* The tumor specific immunoglobulin idiotype on B cell lymphomas provides a suitable target for immunotherapy, and treatment with anti-idiotypic antibodies has achieved therapeutic effects in some patients (Levy and Miller 1991). Similarily, an antibody against a growth factor receptor whose presence on tumor cells is necessary for their proliferation (e.g., IL-2 receptor α-chain on some T cell leukemias) may be therapeutically effective (Waldmann 1991). One of the disadvantages of antibodies seems to be their inability to counteract larger tumor masses and, consistently, Riethmüller et al. (1994) showed that a tumor reactive antibody applied to patients with minimal residual disease prolonged survival.

Combined new strategies
1 + 2: Gene-modified tumor cells, tumor antigen peptide + adjuvants
1 + 3: Tumor antigen peptide loaded APC
2 + 3: CTL clones + IL2
2 + 4: Idiotype–GM-CSF fusion protein;
3 + 4: Gene-modified T cells (Ab-FcRγ); LAK or T cells + bispecific mAb

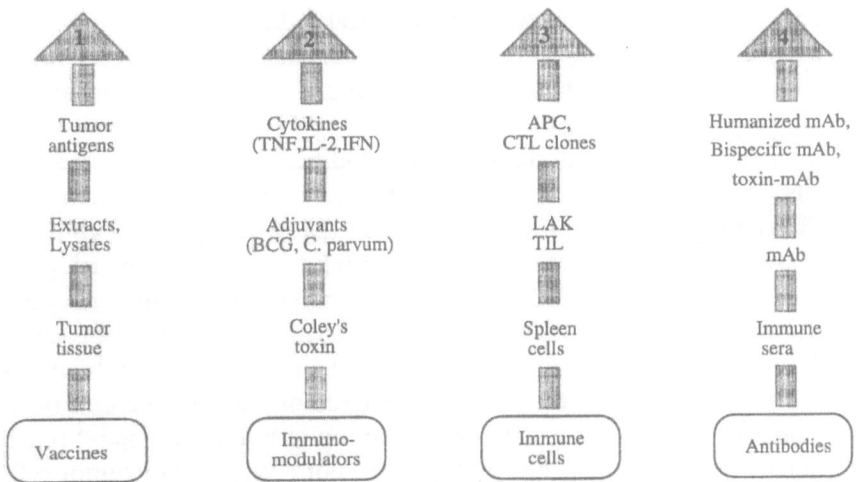

Fig. 1. The development of concepts for cancer immunotherapy. Four general principles for tumor therapy and their development during the last centenary are illustrated. New strategies that combine two different principles and are currently mainly investigated in experimental models include the following: *1 + 2*, gene-modified tumor cells, tumor antigen peptide plus adjuvants; *1 + 3*, tumor antigen peptide-loaded antigen-presenting cells (*APC*); *2 + 3*, cytotoxic T lymphocyte (*CTL*) clones plus interleukin (*IL*)-2; *2 + 4*, idiotype–granulocyte-macrophage colony-stimulating factor (GM-CSF) fusion protein; *3 + 4*, gene-modified T cells (Ab-FcRγ), lymphokine-activated killer cells (*LAK*) or T cells plus bispecific monoclonal antibodies (*mAb*). *TIL*, tumor-infiltrating lymphocytes; *TNF*, tumor necrosis factor; *IFN*, interferon, *BCG*, bacille Calmette-Guérin

Nevertheless, the unsuccessful experience with immunotherapy in cancer patients exceeds by far the positive results (Oettgen and Old 1991; Dalgleish 1994). It seems as if immunotherapy so far was just successful enough to keep the subject alive but was never sufficiently successful for widespread clinical application. Immunotherapy currently under experimental investigation largely encompasses combined strategies (Fig. 1). Most approaches have as common aim either inducing tumor reactive T cells in vivo or generating and expanding tumor specific T cells in vitro for adoptive transfer.

- *Immunomodulators/antibodies.* A fusion protein consisting of an idiotype-specific antibody and granulocyte-macrophage colony-stimulating factor (GM-CSF) but not both molecules individually acted as a vaccine against a lethal challenge with a B cell lymphoma (Tao and Levy 1993). Because GM-CSF stimulates antigen presentation by antigen presenting cells (APC) (Steinmann 1991), the physical linkage of the idiotype antibody to GM-CSF may enhance the uptake and presentation of tumor antigens. It is not yet clear whether cellular or humoral mechanisms are responsible for this effect.

- *Immunomodulators/vaccines.* Genetic modification of tumor cells as discussed below is used for increasing the immunogenicity of tumor cells. This approach relies on the assumption that the respective tumor expresses antigen(s) against which an immune response can be triggered by appropriate stimulation. This approach seems legitimate, because the evidence for this assumption increases, however, tumor (associated or specific) antigens have been cloned only in few cases and the tumor cell often is the only source for putative tumor antigens (Pardoll 1994). Some tumor antigens isolated by the group of Boon et al. (1994) reflect normal cellular gene products expressed not exclusively by tumor cells, such as MAGE-1, MAGE-3, MART-1 and tyrosinase. However, peptides thereof are recognized by cytotoxic T lymphocytes (CTL) in the context of different major histocompatibility complex (MHC) class I molecules. Both strategies (gene modified tumor vaccines and peptide/adjuvants vaccines) mean a continuation of classical tumor cell/adjuvants vaccines but are much more clearly defined.

- *Vaccine/immune cells.* It is well recognized that a key cellular element for induction of specific immunity are APC, e.g., dendritic cells (Steinmann 1991). Culture conditions for in vitro expansion of dendritic cells have been improved and tumor antigen pulsed dendritic cells upon adoptive transfer have been shown to possess vaccine character in a mouse B cell lymphoma model (Flamand et al. 1994). The protective effect against the parental tumor was mediated by antibodies.

- *Immune cells/immunomodulators.* In experimental tumor models the efficacy of adoptively transferred T cells to eliminate established tumors has been shown in several cases (Greenberg 1991; Melief 1992). Experimental settings which used either tumor specific CTL clones or spleen cells from tumor immunized mice were most successful. The moderate therapeutical effect of human tumor infiltrating lymphocytes (e.g., from melanomas) after in vitro expansion and adoptive transfer along with IL-2 can presumably be explained by the usually unknown number of *specific* CTL. Virus (e.g., cytomegalovirus, CMV; Epstein-Barr virus, EBV; human immunodeficiency virus, HIV) specific CTL clones are currently tested in patients undergoing bone marrow transplantation (Quinnan et al. 1982; Riddel et al. 1992).

– *Immune cells/antibodies.* Several experimental strategies are followed to direct T cells or lymphokine activated killer (LAK) cells more specifically to tumors. Coinjection of LAK cells together with a tumor reactive antibody into SCID mice bearing a human colon tumor inhibited the tumor growth, that was presumably mediated by antibody dependent cell mediated cytotoxicity (Takahashi et al. 1993). Similarily, the combination of human T cells and a chimeric bispecific antibody recognizing the CD3 as well as CD30 molecules led to destruction of a CD30[+] lymphoma in SCID mice (Renner et al. 1994). Alternatively, T lymphocytes have been transfected with a fusion gene consisting of an antibody variable region with specificity for an ovarian carcinoma and the Fc receptor γ-chain mediating T cell signaling. The resultant cells specifically lysed the tumor cells and secreted cytokine upon antigenic stimulation (Hwu et al. 1993).

This brief overview shows that the development in genetic engineering and cell culture conditions has opened a door for a variety of new approaches for cancer immunotherapy. How they compare to each other in terms of efficacy is largely unknown even in animal models and sufficient clinical experience does not yet exist for any of these strategies.

3
Gene Transfer Methods

Gene transfer systems for eukaryotic cells include both nonviral and viral methods (for a review, see Mulligan 1993). For gene therapy of genetic defects, an ideal gene transfer system should combine several characteristics: high efficiency for in vivo gene transfer, cell type specificity, regulated and stable gene expression, the gene integration site should be either defined or as long-term goal, the defective gene should be replaced by a functional copy through homologous recombination. Additionally, the problem of immunogenicity of exogeneous gene products should be resolved. For construction of tumor cell vaccines the requirements may be less critical.

The gene transfer methods that have been developed up to date can be evaluated by several criteria: in vivo or ex vivo gene transfer, stable or transient gene expression, level of expression of the "therapeutic" gene and gene transfer efficacy (Table 2). Tools mostly used for gene transfer into mouse tumor cells are replication deficient retroviruses, electroporation, liposomes, receptor (transferrin) mediated gene transfer, and more recently recombinant adenoviruses or vaccinia viruses. Current clinical trials with cytokine gene modified tumor cells employ in most cases retroviral gene transfer ex vivo. The high gene transfer efficiency of retroviruses may prevent the in vitro selection of tumor cells that could lead to the loss of tumor antigens and reduces the time required for production of tumor vaccines (e.g., in autologous settings, see below). Additionally, retroviruses stably integrate into the host's genome ensuring a sustained expression. Transient expression of the gene of interest may be sufficient in

Table 2. Gene transfer methods

Method	Efficacy (ex vivo)	Application (in vivo)	Expression	Immunogenicity
Nonviral				
CaPO$_4$ precipitation	±	No	S	–
Electroporation	±	No	S	–
Liposomes	±	Yes	T	No
Ligand–receptor/				
DNAcomplex	+	No	T	–
Direct DNA injection	–	Yes	T	No
Viral				
Retrovirus	++	Yes	S	Yes
Adenovirus	+++	Yes	T	Yes
Adeno-associated virus	++(?)	?	S(?)	?
Vaccinia virus	?	Yes	T	Yes
Herpesvirus	?	Yes	?	Yes

S, stable; T, transient.

some cases, but with stable expression it is easier to standardize the procedure. Clearly, more efficient and cell type specific in vivo gene transfer methods would facilitate many therapeutical approaches.

4
Cytokine Gene-Modified Tumor Cells

Cytokines play an important role for immunotherapy. The number of cytokines constantly increases and if chemokines, colony stimulating factors and angiogenic factors are included, it presumably exceeds 50. Some characteristics of cytokines are listed below:
- Regulate immune response, inflammation, and hematopoesis
- can be expressed by various cell types
- May have different activities on various cell types (differentiation, proliferation, activation, suppression)
- May have similar activity with respect to each other
- Seem to provide primarily a short range signaling between immune and/or nonimmune cells
- Comprise a regulatory network in vivo
- Are usually toxic when systemically applied in unphysiologically large amounts
- Usually have a short half-life in blood
- With few exceptions are not directly cytotoxic or cytostatic for tumor cells
- Show some therapeutical benefit upon systemic application to cancer patients in selected situations (e.g., IL-2 in some melanoma or renal cell carcinoma patients)

Basically, cytokine gene modified tumor cells have been analyzed in two ways: (1) By analysis of the tumorigenicity of the transfected tumor cells themselves and (2) by the analysis of such cells to act as a vaccine. The experimental setting for the latter analysis consisted in either pre-immunization of mice with the cytokine gene modified tumor cells and subsequent contralateral challenge with the parental tumor cells or, in a therapeutic model, injection of the parental tumor cells and, shortly afterwards, treatment of the mice with the gene modified cells. Read-out system was either recording the survival rate of animals without tumor or monitoring the inhibition of metastasis during a certain period after immunization. In most cases but not always (e.g., GM-CSF, see below) loss of tumorigenicity of the transfected tumor cells and their vaccine effects seem to correlate.

4.1
Many But Not All Cytokines Reduce Tumorigenicity When Expressed by Transfected Tumor Cells

Many cytokines have been expressed in different rodent tumor cells and presumably there are some more studies which we may have overlooked (Table 3). Most experiments were done with cytokines IL-2, IL-4, IL-6, IL-7, IL-10, tumor necrosis factor (TNF), IFNγ, and GM-CSF. The classification *decreased* or *unchanged* tumorigenicity of the transfected tumor cells has to be somewhat arbitrary because of different experimental designs, differences in the amount of cytokine expression and different criteria for judgement of changed immunogenicity/tumorigenicity. *Decreased* tumorigenicity is defined by a strongly reduced tumorigenicity of the cytokine gene transfected tumor cells, e.g., a high percentage of the mice completely rejected an otherwise lethal tumor cell challenge. *Unchanged* tumorigenicity means that expression of the transfected cytokine only slightly or not at all reduced tumor growth of the respective tumor cells. With few exceptions (e.g., IL-6 expression by 3LL cells or TNF expression by MCA-102 cells) secretion of the transfected cytokine did not change the duplication time of the cells in vitro. By (most often subcutaneous) injection of the transfected cells into mice it was shown that IL-1, IL-2, IL-3, IL-4, IL-6, IL-7, IL-10, IL-12, IL-13, TNF, LT, IFNα, IFNγ, G-CSF, MCAF and IP-10 suppressed tumor growth in at least one tumor model (for references, see Table 3). In several models it has been shown that parallel application of neutralizing antibodies specific for the transfected cytokine restored tumorigenicity, thus attributing reduction or loss of tumorigenicity directly to the transfected cytokine (e.g., Tepper et al. 1989). Most reliably, IL-2 expression by the tumor cells led to their rejection in (syngeneic) mice, followed by IL-7 and IL-4. Several cytokines (IL-1, IL-4, IL-6, IL-10, TNF, IFNγ, MCAF) suppressed tumor growth in some but completely failed or only marginaly reduced tumorigenicity in other tumor models (IL-1 in MBT-2, Saito et al. 1994; IL-4 in B16-F10, Dranoff et al. 1993; IL-6 in J558L and TS/A, Blankenstein et al. 1991a, Allione et al. 1994; IL-10 in TS/A cells, unpublished observation; TNF in ESB and MCA-102, Qin et al. 1993, Karp et al. 1993; IFNγ in B16-F10 and TS/A, Dranoff et al. 1993, Lollini et al. 1993; MCAF in B16 cells, Bottazzi et al. 1992). Some cytokines (IL-5,

Table 3. Cytokine gene-transfected rodent tumors

Gene transfected	Tumorigenicity of the indicated cell lines		
	Decreased	Unchanged	References
IL-1	Oncogene transformed NIH3T3		Douvdevani et al. 1992
			Saito et al. 1994
IL-2	Rat-1	MBT-2	Bubenik et al. 1988
	X63		Bubenik et al. 1990
	CT-26, B16-F10		Fearon et al. 1990
	CMS-5		Gansbacher et al. 1990a
	HSLNV		Russell et al. 1991
	P815		Ley et al. 1991
	TS/A		Cavallo et al. 1992
	J558L		Hock et al. 1993a
	MCA-102		Karp et al. 1993
	3LL		Porgador et al. 1993b
			Ohe et al. 1993
	MBT-2		Connor et al. 1993
	4T07		Tsai et al. 1993
	B16-F10		Dranoff et al. 1993
	EL4		Visseren et al. 1994
	Neuro-2a		Katsanis et al. 1994
IL-3	Line 1		Pulaski et al. 1993
	FSA		McBride et al. 1994
IL-4	J558L, K485,		Tepper et al. 1989, 1992
	J558L		Li et al. 1990
			Hock et al. 1993a
	Renca		Golumbek et al. 1991
	CHO		Platzer et al. 1992
		B16-F10	Dranoff et al. 1993
	3LL		Ohe et al. 1993
	EB		Khazaie et al. 1994
	TS/A		Pericle et al. 1994
IL-5		J558L, TS/A	Krüger-Krasagakes et al. 1993
		B16-F10	Dranoff et al. 1993
IL-6		J558L	Blankenstein et al. 1991a
	3LL[a]		Porgador et al. 1992
	MCA-205, MCA-207		Mullen et al. 1992
		B16-F10	Dranoff et al. 1993
		TS/A	Allione et al. 1994

Table 3. Continued

Gene transfected	Tumorigenicity of the indicated cell lines		
	Decreased	Unchanged	References
IL-7	J558L, TS/A		Hock et al. 1991, 1993a
	FSA		McBride et al. 1992
	203-glioma		Aoki et al. 1992
IL-10	CHO		Richter et al. 1993
		TS/A	unpublished
	CL8-1		Suzuki et al. 1995
	TS/A		Giovarelli et al. 1995
	B16-F1		Gérard et al. 1996
IL-12		C-26	Martinotti et al. 1995
	MCA-207, MCA-102		Tahara et al. 1995
IL-13	P815		Lebel-Binay et al. 1995
TNF	CHO		Oliff et al. 1987
			Qin et al. 1995
	J558L		Blankenstein et al. 1991a
			Hock et al. 1993a
	1591-RE		Teng et al. 1991
	MCA-205		Asher et al. 1991
	EB	ESB[b]	Qin et al. 1993
		MCA-102[a]	Karp et al. 1993
		B16-F10	Dranoff et al. 1993
	TS/A		Allione et al. 1994
LT	CHO		Qin et al. 1995
	J558L		Qin and Blankenstein 1995
IFN-α	FLC		Ferrantini et al. 1993, 1994
IFN-γ	C1300		Watanabe et al. 1989
	CMS-5		Gansbacher et al. 1990b
	SP1, CT-26		Esumi et al. 1991
		MCA-101	Restifo et al. 1992
	J558L		Hock et al. 1993a
	MBT-2		Connor et al. 1993
	3LL		Porgador et al. 1993a
		B16-F10	Dranoff et al. 1993
		TS/A[b]	Lollini et al. 1993
		MBT-2	Saito et al. 1994
M-CSF		J558L	Dorsch et al. 1993

Table 3. Continued

Gene transfected	Tumorigenicity of the indicated cell lines		
	Decreased	Unchanged	References
G-CSF	C-26		Colombo et al. 1991
			Stoppacciaro et al. 1993
GM-CSF		B16-F10[c]	Dranoff et al. 1993
		MBT-2[c]	Saito et al. 1994
		TS/A	Allione et al. 1994
		J558L	Qin et al. 1996
MCAF	CHO		Rollins and Sunday, 1991
		B16	Bottazzi et al. 1992
IP-10	J558L, K485		Luster and Leder 1993

The classification 'decreased' or 'unchanged' tumorigenicity had to be somewhat arbitrary, because different criteria were chosen by different investigators. 'Unchanged' tumorigenicity means that the transfected cytokine decreased tumorigenicity of the transfected tumor cells only slighly or not at all; it does not take into account the fact that higher expression levels of the cytokine or variation in the number of injected tumor cells might lead to different results. 'Decreased' tumorigenicity means that the tumorigenicity of the transfected cells was strongly diminished, e.g., at least some of the mice completely rejected the tumor cells.
IL, interleukin; TNF, tumor necrosis factor; IFN, interferon; LT, lymphotoxin; GM-CSF, granulocyte-macrophage colony-stimulating factor; CHO, Chinese hamster ovary; MCAF, monocyte chemotaxis activating factor.
[a]Doubling time of the cells in vitro was delayed.
[b]Metastatic potential of the cell was increased.
[c]GM-CSF in tumor cells B16-F10 and MBT-2 differed from the other cytokines, because even though the transfected cells grew progressively as tumors, immunization with irradiated GM-CSF-secreting cells induced tumor immunity.

M-CSF) so far failed to reduce tumorigenicity of respectively transfected cells even though infiltrating immune cells were detected in the tumor (Krger-Krasagakes et al. 1993; Dorsch et al. 1993). GM-CSF secreting tumor cells grew progressively as tumor. However, GM-CSF seems to differ from other cytokines because immunization with irradiated cells induced tumor immunity in several tumor models (Dranoff et al. 1993; Saito et al. 1994) (see below).

It is important to point out that the amount of cytokine produced by the tumor cells often correlated with the decrease of tumorigenicity. Whether different results with the same cytokine in different tumor models is due to differences in cytokine expression is currently not known.

4.2
Direct and Indirect Effects
of Transfected Cytokines on Tumor Growth

As already mentioned, expression of a cytokine by tumor cells in most cases does not change their growth characteristics in vitro which suggests that loss of tumorigenicity results from interaction between the cytokine transfected tumor cells and the host. However, phenotypical changes other than the alteration of the generation time of tumor cells are likely to occur via an autocrine loop provided that the tumor cells express the respective cytokine receptor. Several cytokine receptors such as IL-4, IL-6, TNF, and IFNγ receptors are widely distributed on numerous cell types. The clearest example for a direct effect of the transfected cytokine on the tumor cell phenotype is induction of MHC class I expression by IFNγ. In most of the examples shown in Table 3, endogenous IFNγ expression up-regulated MHC class I molecules and potentially antigen presentation but this was not necessarily connected with reduced tumorigenicity. Whether different cytokine receptors are expressed by the tumor cells listed in Table 3 has not been analyzed so far. Other examples of direct effects of the transfected cytokine on the tumor cell phenotype are the inhibition of TGF-β after IL-7 gene transfer in a human melanoma cell line (Miller et al. 1993) or increased IL-6 expression by CHO cells following TNF or LT gene transfer (Qin et al. 1995). It is likely that other phenotypical changes have gone unnoticed.

Another possibility of inducing phenotypical changes of tumor cells comes from a different source, namely the expression of other coding sequences in the expression vector. This could lead not only to a change in immunogenicity of the tumor cells, but also to a change in antigenicity which is known as 'xenogenization'. If the coincidentally expressed antigens are weak, this per se must not reduce the tumorigenicity of the 'mock'-transfected control cells, but possibly can contribute to loss of tumorigenicity when there is the additional expression of a cytokine.

4.3
Immunological Mechanisms of Tumor Rejection

The loss of tumorigenicity of the variously cytokine gene transfected tumor cells results from an increased immunogenicity: after injection into mice the tumor cells start to grow and concomitantly secrete the transfected cytokine. This can induce an inflammatory reaction whereby – depending on the cytokine (and the investigated tumor) – different immune cells are attracted, activated and eventually destroy the tumor cells. The immunological mechanisms leading to rejection of the cytokine gene transfected tumor cells have partially been resolved and based on the following types of experiments:

1. Immune cells were detected in the tumor tissue by immunohistochemistry or electron microscopy. The detection of certain cell types in the cytokine producing but not the parental or mock-transfected tumor is suggestive for participation of these cells in the antitumor response. Indeed, sufficient examples exist which confirmed the antitumor activity of immune cells present in the tumor by parallel cell depletion studies. As an example, $CD4^+$ and $CD8^+$ T cells specifically infiltrated J558-IL7 tumors and both cell types were mandatory for complete tumor rejection (Hock et al. 1991, 1993a). On the other hand, immune cells can infiltrate the tumor without beeing involved in the antitumor response (e.g., $CD4^+$ cells in J558-TNF tumors, Hock et al. 1993a). Moreover, despite of beeing heavily infiltrated by immune cells, cytokine producing tumors can progressively grow. This, for example, has been shown for macrophages in M-CSF and eosinophils in IL-5 producing tumors (Dorsch et al. 1993, Krüger-Krasagakes 1993). An inverse correlation between loss of tumorigenicity and presence of a cellular infiltrate was found with IL-10 gene transfected CHO cells in immunodeficient mice (Richter et al. 1993). Parental but not IL-10 secreting tumors, which were unable to grow, were infiltrated by macrophages.

2. Tumorigenicity of the transfected cells was in parallel analyzed in syngeneic immunocompetent and immunodeficient mice (e.g., T cell deficient nude or T and B cell deficient SCID). This has shown that for a variety of cytokines, both T and non-T cells contributed to the tumor suppressive effect (Fig. 2). In the absence of T cells tumor growth was effectively suppressed by a number of cytokines such as IL-2, IL-4, TNF, or IFNγ. However, this effect was often transient and most of the mice developed a tumor after a long latency period due to the lack of

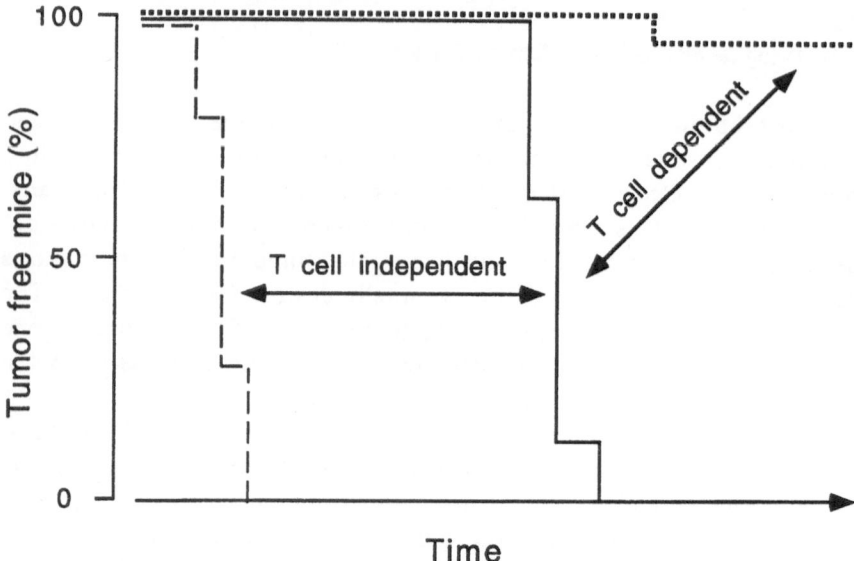

Fig. 2. Biphasic rejection mechanism of cytokine-gene modified tumors observed with J558L cells transfected to produce interleukin (IL)-2, IL-4, tumor necrosis factor (TNF), and interferon (IFN)-γ. The respectively cytokine-secreting cells were injected into T cell-deficient nude mice (*solid line*) or syngeneic BALB/c mice (*dotted line*) and tumor growth was analyzed. As a control, parental cells were also analyzed; they grow with very similar kinetics in both mouse strains (*dashed line*). Results are described in detail in Hock et al. (1993a)

T cells (Hock et al. 1993a). It is important to note that tumors which had grown in nude mice after a latency period in case of IL-2, IL-4 and TNF, but for unknown reasons not in case of IFNγ always had stopped or strongly reduced expression of the transfected cytokine. This demonstrates that one function of the T cells is to prevent the outgrowth of cytokine loss variants. However, almost always few nude mice completely rejected the cytokine gene transfected tumor and conversely, some syngeneic mice developed a tumor after a long latency (Hock et al. 1993a). The type of T cells required for complete tumor rejection cannot be determined by this assay.

3. Tumorigenicity was analyzed in mice in which defined cell populations were selectively eliminated by appropriate antibodies. This assay most clearly attributes the antitumor response against cytokine gene transfected tumors to a particular cell type and has been done for T cell

subsets CD4$^+$ and CD8$^+$, Mac1$^+$ cells (predominantly macrophages), granulocytes (neutrophilic or eosiniphilic) and asialo GM1$^+$ cells (NK cells). However, this leaves the question open whether the respective cell population is always the direct target for the transfected cytokine. We have previously found that, regardless which of the cytokines IL-2, IL-4, IL-7, TNF and IFNγ was produced by the same tumor, CD8$^+$ T cells were always needed for long-term tumor elimination even though tumor growth was quite effectively suppressed for a certain period in CD8$^+$ T cell depleted mice. Therefore, we have argued that the non-specific inflammatory cells prevent rapid tumor burden without being able to completely eliminate the tumor cells in most cases but giving CD8$^+$ T cells the chance to develop cytotoxic activity which, however, must not necessarily be stimulated in all cases by the transfected cytokine (Hock et al. 1993a).

4. CTL activity of spleen cells of immunized mice was analyzed. This assay indirectly measures systemic tumor immunity. In several cases the immune status of mice having rejected the cytokine gene transfected tumor and CTL activity correlated with each other (see for example Fearon et al. 1990, Gansbacher et al. 1990a). However, in vitro cytolytic activity and in vivo antitumor activity do not always correlate (Barth et al. 1991; Pericle et al. 1994).

In general, the contribution of a certain cell type to tumor suppression must not always result from a direct effect of the transfected cytokine on these cells but rather can be due to interaction between different immune cells. The presence of macrophages and eosinophilic granulocytes in IL-7 producing tumors was dependent on CD4$^+$ T cells, and tumor growth was restored when CD4$^+$ T cells were depleted in the mice (Hock et al. 1991, 1993a).

Regarding these quite different assays and, more importantly, the fact that results from different tumor models were not always consistent, it becomes clear that any conclusion has to remain tentative. Figure 3 tries to summarize the immunological mechanisms of cytokine dependent tumor rejection. More or less all possible immune cells have been detected in the different cytokine producing tumors, many cytokines simultaneously involve several cell types, requirement for non-T cells and T cells is a common observation, several cytokines at least partially activate similar effector mechanisms, and inducer and effector cells are often difficult to distinguish from each other in this experimental setting. Several reviews

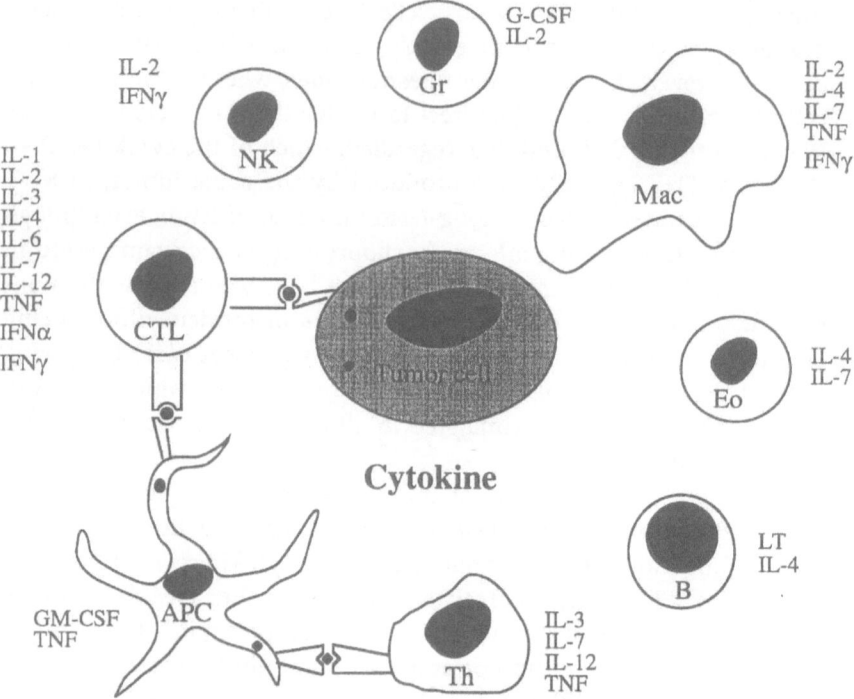

Fig. 3. Host immune cells that may be activated by cytokine gene-transfected tumor cells. Various types of host immune cells have been detected in tumors that produce cytokines upon gene transfer. Major histocompatibility complex (*MHC*)-restricted antigen presentation by antigen-presenting cells (*APC*) or MHC class I$^+$ tumor cells is also illustrated. *NK*, natural killer cells; *Gr*, neutrophilic granulocytes; *Mac*, macrophages; *Eo*, eosinophils; *B*, B lymphocytes; *Th*, helper T lymphocytes. *IL*, interleukin; *TNF*, tumor necrosis factor; *IFN*, interferon; *GM-CSF*, granulocyte-macrophage colony-stimulating factor; *CTL*, cytotoxic T lymphocyte

addressing the immunological mechanisms of rejection of cytokine gene transfected tumors have been published (Blankenstein et al. 1991b; Colombo et al. 1992; Pardoll 1993; Blankenstein 1994).

4.4
Influence of the Tumor Model on Cytokine Effects

The analysis of antitumor effects with cytokine gene transfected tumor cells includes a number of different tumor cell types (Table 4). This shows

Table 4. Transfected cytokines suppress in vivo growth of a variety of transplanted tumor cells

Type of tumor	Cell line
Fibrosarcoma	CMS-5, FSA, MCA-102, MCA-205, MCA-207
Colon carcinoma	CT-26, C-26
Plasmacytoma	J558L, X63
Adenocarcinoma	K485, TS/A, SP1
Renal cell carcinoma	Renca
Bladder carcinoma	MBT-2
Lung carcinoma	3LL, line 1
Glioblastoma	203-glioma
Neuroblastoma	C1300, neuro-2a
Melanoma	B16(-F10)
T lymphoma	EB, EL4
Mastocytoma	P815
Skin tumor	1591-RE
Fibroblast	Oncogene transformed NIH3T3
Friend leukemia	FLC
Rat sarcoma	HSNLV
Hamster ovarian carcinoma	CHO

Cytokines which have been used are shown in Table 3. Most of the tumors were induced by chemical carcinogens, ultraviolet (UV) radiation, transfected oncogenes or virus. According to the literature, tumors B16, Renca, C1300, SP1, and TS/A arose spontaneously.

that the strong antitumor effect of local cytokines is applicable to most if not all tumor cell types. However, different tumor cells, even if of the same histological origin, may yield different or even opposite results when transfected with the same cytokine. This is illustrated by some selected examples:

- *Susceptibility/resistance to a given cytokine.* TNF expression by EB cells led to their rapid rejection whereas similar amounts of TNF secreted by ESB cells did not reduce tumorigenicity. Similarily, TNF gene transfected MCA-205 but not MCA-102 cells lost their tumorigenicity. IL-6 expression by 3LL, MCA-205 and MCA-207 cells but not by J558L, B16-F10 and TS/A cells diminished their tumorigenicity. IL-10 expression by CHO cells abolished their tumor growth even in nude or SCID mice, similar levels of IL-10 expressed by TS/A cells did not change their tumorigenicity at all (unpublished observation). IFNγ gene transfected tumor cells were variably rejected or progressively grew as tumor dependent on the tumor model but obviously independent on the amount of IFNγ production (for references and further examples, see Table 3).

– *Differences in the immunological antitumor mechanisms.* The non-T cell infiltrate in J558-IL2 tumors mainly consisted of macrophages (Hock et al. 1993a), that in TS/A-IL2 tumors of neutrophilic granulocytes (Cavallo et al. 1992). Furthermore, NK cells contributed to destruction of some (Hock et al. 1993a, Karp et al. 1993) but not other IL-2 producing tumors (Cavallo et al. 1992). The rejection of J558-IL7 cells was completely dependent on CD4$^+$ T cells (Hock et al. 1991), but CD4$^+$ T cells were not needed for rejection of IL-7 producing 203-glioma cells (Aoki et al. 1992). Loss of tumorigenicity of TNF producing tumors could be reversed by CD4$^+$ T cell depletion in one (Asher et al. 1991), but not in another tumor model (Hock et al. 1993a). NK cells contributed to tumor suppression in the J558L (Hock et al. 1993a) but not in the C1300 tumor model (Watanabe et al. 1989) when the cells expressed IFNγ.

Differences in cell surface molecule expression, endogenous cytokine production, tumor growth kinetics, susceptibility to effector mechanisms, inherent tumor immunogenicity, the mouse strain or as mentioned above the level of expression of the transfected cytokine may contribute to these discrepancies.

4.5
Enhancement of Tumorigenicity of Cytokine Gene-Transfected Cells

Increased tumorigenicity following transfer and expression of cytokine genes has been demonstrated in several models and so far may result from three principles:

1. *Autocrine stimulation.* Several immortalized cell lines whose growth in culture was strictly dependent on exogenous cytokines were subjected to gene transfer of the respective cytokine genes. Through an autocrine pathway the cells could stimulate their proliferation via the transfected cytokine and the exogenous cytokine became dispensable. With cytokines IL-2, IL-3, IL-5, IL-6, IL-9 and GM-CSF it has been shown that such autonomously growing cells had achieved tumorigenic potential whereas the parental, cytokine-dependently growing or the mock-

transfected cells had not. The cell lines were all of hematopoietic origin. It should be noted that in some models small amounts of the cytokine (e.g., IL-2) was produced which may have been sufficient for autocrine stimulation but not sufficient for induction of an antitumor response (for a review, see Blankenstein et al. 1991b).

2. *Immune suppression.* TGF-β, which among other activities has inhibitory effects on activation and proliferation of CTL, has been expressed in a highly immunogenic tumor cell line. Unlike the parental cells the TGF-β producing cells, even though still lysed by CTL in vitro, poorly induced CTL in vivo and grew as tumor in transiently immunosuppressed mice (Torre-Amione et al. 1990). Recently, we have shown that IL-10 spontaneously released by J558L tumor cells inhibited dendritic cell infiltration and suppressed an immune response induced by GM-CSF which was transfected into this tumor line (Qin et al. 1996).

3. *Augmented metastasis.* As discussed already, TNF and IFNγ are both able to induce tumor rejection in several models when expressed by the tumor cells after gene transfer. Both molecules, however, may have opposite effects. Expression of TNF in the metastatic cell line ESB instead of suppressing tumor growth augmented spontaneous metastasis formation in the liver (Qin et al. 1993). TNF upregulates a variety of adhesion molecules on endothelial cells and promotes transendothelial migration of certain cell types. ESB cells express ligands for these adhesion molecules which could facilitate arrest in the bloodstream and enable ESB-derived TNF to allow stabilization of this interaction by upregulation of adhesion molecules on endothelial cells. Furthermore, TNF may stimulate transendothelial migration of ESB. TS/A cells transfected to produce IFNγ generated more lung metastases upon intravenous injection into mice compared to the parental cells (Lollini et al. 1993). This effect was more pronounced with low IFNγ producing cells which had upregulated MHC class I expression and might be less sensitive to NK cell mediated lysis (Taniguchi et al. 1987).

The effects observed with cytokine gene transfected tumor cells are summarized in Table 5. Cytokines can either suppress tumor growth, promote tumor growth or leave the tumorigenicity of the cells unchanged. Tumor suppression results (with the possible exception of IL-10; Richter et al. 1993) from induction of an inflammatory reaction and activation of the host's immune system. Tumor promotion can be due to autocrine stimulation, local immune suppression or augmentation of metastasis.

Table 5. Effects observed with cytokine gene-transfected tumors

Effect	Cytokine gene
Tumor suppressive	
Activation of host immunity	IL-1, IL-2, IL-3, IL-4, IL-6, IL-7, IL-12, IL-13, TNF, LT, IFNα, IFNγ, G-CSF, GM-CSF, MCAF, IP-10
Interruption of tumor–host interaction	IL-10
Tumor promoting	
Autocrine stimulation	IL-2, IL-3, IL-5, IL-6, IL-9, GM-CSF
Immune suppression	TGF-β, IL-10
Augmented metastases	TNF, IFNγ
Tumorigenicity not changed	
IL-1, IL-4, IL-5, IL-6, IL-10, TNF, IFNγ, M-CSF, GM-CSF, MCAF	

IL, interleukin; TNF, tumor necrosis factor; LT, lymphotoxin; IFN, interferon; GM-CSF, granulocyte-macrophage colony-stimulating factor; MCAF, monocyte chemotaxis activating factor; TGF, transforming growth factor.

5
Spontaneous Cytokine Expression by Tumor Cells

The main reason for using tumor cells as basis for vaccine studies is that they serve in most cases as the only source for (putative) tumor antigens. Regardless whether immune surveillance against tumor cells exists or tumor cells present true tumor specific antigens as in the case of virus induced cancer, the evidence from mouse and human tumors increases that a host immune response can be activated against many if not all tumors. However, when one thinks about introducing a cyokine gene into tumor cells, one has to be aware of the fact that tumor cells can – and in fact do – constitutively produce a variety of cytokines. A literature research showed that almost all cytokines can be expressed by some human tumors (Table 6). Certainly, this is not a complete list. It includes results of cytokine detection in supernatant of cultured cells by enzyme-linked immunosorbent assay (ELISA) or bioassay, detection of cytokine mRNA in tumor biopsies by reverse transcription–polymerase chain re-

Table 6. Endogenous cytokine expression by human tumor cells

Cytokine	Cell
IL-1	melanoma, fibrosarcoma, MM, bladder carcinoma, thymoma, HL, T cell leukemia, ALL, AML, CML, CLL, hepatoblastoma, glioblastoma, ovarian cancer, thyroid cancer, LC, squamous cell carcinoma
IL-2	melanoma, HTLVI$^+$ T cell leukemia
IL-3	ALL, mastocytoma
IL-4	HL, BCC
IL-5	HL, NHL, BCC
IL-6	RCC, MM, melanoma, osteosarcoma, peritoneal mesothelioma, cardiac myxoma, glioblastoma, neuroblastoma, pituitary adenoma, hepatoma, AML, ALL, CLL, HL, NHL, LCAL, Burkitt lymphoma, LC, ovarian carcinoma, thyroid carcinoma, bladder carcinoma, prostate carcinoma, colon carcinoma, breast carcinoma, Karposi sarcoma
IL-7	CLL
IL-8	melanoma, ovarian carcinoma, hepatocellular carcinoma, astrocytoma, lung giant cell carcinoma, AML, CLL, HL
IL-9	HL, LCAL
IL-10	melanoma, RCC, colon carcinoma, neuroblastoma, BCC, ovarian carcinoma, EBV$^+$ B lymphoma
TNF	cervical carcinoma, RCC, melanoma, astrocytoma, glioblastoma, ovarian carcinoma, HL, NHL, CLL, AML, CML
LT	melanoma, sarcoma, NHL, HL
IFNγ	melanoma, ovarian carcinoma, astrocytoma
G-CSF	bladder carcinoma, multiple myeloma, LC, astrocytoma, thyroid cancer, glioblastoma, breast sarcoma, hepatocellular carcinoma, HL, NHL, CLL
GM-CSF	melanoma, colon carcinoma, LC, glioblastoma, astrocytoma, BCC, ovarian carcinoma, thyroid cancer, HL, AML, CLL, CML
TGF	pancreatic carcinoma, melanoma, breast cancer, LC, glioblastoma, astrocytoma, colon cancer, ovarian carcinoma, hepatocellular carcinoma, MM, HL

multiple myeloma (MM), Hodgkin lymphoma (HL), acute lymphoblastic leukemia (ALL), acute myeloid leukemia (AML), chronic myeloid leukemia (CML), chronic lymphocytic leukemia (CLL), lung carcinoma (LC), human T-cell leukemia virus type I (HTLVI), basal cell carcinoma (BCC), non Hodgkin lymphoma (NHL), renal cell carcinoma (RCC), large cell anaplastic lymphoma (LCAL) action (RT-PCR) or in situ hybridisation, or detection of augmented cytokine levels in serum or ascites of cancer patients. In some cases, it has to be demonstrated that the tumor cells also secrete biologically active cytokine in vivo and the cytokine is indeed secreted by tumor cells rather than by tumor infiltrating cells. The reasons for cytokine expression by tumor cells in most cases is not known. Some cytokines (e.g., IL-2, IL-3, IL-4, IL-5, IL-7, IL-9, IFNγ) seem to be expressed exceptionally and predominantly in hematopoietic malignancies. This could be due to their restricted expression pattern in non-malignant cells and, if the histological cell type from which the tumor cells is derived, is the natural source for the cytokine, expression by the tumor cells could be coincidental. In some cases simultaneous expression of the cytokine and the respective cytokine receptor may support the proliferation of tumor cells by an autocrine mechanism (see above and Sporn and Roberts 1985; Lang and Burgess 1990). Some cytokines (e.g., IL-1, IL-6) are frequently expressed by tumor cells. Again, this can reflect the broad expression of these cytokines by different cell types and must not necessarily be causally related to the malignant phenotype. For cytokines such as TGF-β and IL-10, which have immune suppressive activities in several experimental models, it may not be surprising that they are expressed by a variety of different human tumors. For both cytokines it has been shown that expression by mouse tumor cells can strongly diminish a tumor specific immune response (Torre-Amione et al. 1990; Qin et al. 1996). G-CSF and GM-CSF are expressed by a number of tumor cells. Physiologically, few cell types produce these cytokines. Recently, the granulocyte dependent growth of a mouse tumor in nude mice has been shown (Pekarek et al. 1995) raising the question whether tumor cell secreted G-CSF (or any other cytokine) induce granulocytes (or other host cells) for the support of tumor growth. The hypothesis that tumor associated macrophages induced by tumor cytokines may promote tumor growth has been put forward by Mantovani et al. (1992). As an alternative to tumor growth promoting activities, cytokines constitutively expressed by tumor cells may explain the immunogenicity of some tumors, e.g., some renal cell carcinoma or melanomas.

What has become obvious from this incomplete survey of cytokines spontaneously expressed by tumor cells is that many of the cytokines which upon gene transfer suppress tumor growth in the one or the other tumor model are the same cytokines which can also be expressed by progressively growing human tumors. This must not necessarily be in contradiction to each other, because often the amount of transfected cytokines needed to obtain tumor suppressive effects was orders of magnitude higher than that spontaneously released by tumor cells. This leaves us with some unpleasant questions: (a) What is the amount of cytokine required to induce a biological effect in vivo? (b) In which way does the level of cytokine released by the tumor cells determine the type of immune response? IL-10 may serve as a good example. Spontaneous expression of 5 ng/ml IL-10 by J558L cells clearly inhibited an immune response (Qin et al. 1996), whereas tumor cells transfected to produce 50- to 100-fold more IL-10 were rejected (Suzuki et al. 1995). (c) Which of the effects observed with cytokine gene transfected tumors in vivo is a true cytokine-specific effect rather than reflects inflammation due to the unphysiological excess amounts of cytokine present at the tumor site?

Together, tumor cells can constitutively express a number of cytokines, different specimens within one histological tumor cell type may vary extremely in their cytokine expression pattern, and immunological effects induced by transfected cytokines will presumably be influenced by other, tumor cell derived, cytokines. Such cytokines may be expressed coincidentally by the tumor, may support tumor growth in one or another way, or may (unwanted by the tumor cell) stimulate a host immune response. Other phenotypical properties of the tumor cells, first of all expression of major histocompatibility complex molecules (Bernards 1987), or adhesion molecules most likely will also contribute to the outcome of gene modified tumor cell vaccines.

6
Expression of Cell Surface Molecules to Increase Tumor Immunogenicity

6.1
Major Histocompatibility Complex Molecule Expression by Tumor Cells

An alternative strategy to increase the immunogenicity of tumor cells is expression of cell surface molecules on tumor cells. Examples have been provided for de novo expression of MHC class I and class II, B7, ICAM-1 and CD2 molecules. Expression of syngeneic MHC class I molecules on tumor cells which is thought to restore antigen presentation to CD8[+] CTL in some (Hui et al. 1984, Wallich et al. 1985, Tanaka et al. 1985) but not all cases leads to tumor rejection (Kärre et al. 1986). Tumor cells transfected to express an allogeneic MHC class I molecule were rejected and induced protective immunity against the wild-type tumor (Itaya et al. 1987). This effect may be related to an 'immunological bystander' mechanism whereby an immune response against a strong antigen provides help for induction of an immune response against a weak antigen. This result is important when allogeneic tumor cell vaccines are considered. Several mouse tumor cells have been transfected with syngeneic or allogeneic MHC class II genes (Ostrand-Rosenberg et al. 1990; James et al. 1991; Leach and Callahan 1995). Consistently, the cells were more immunogenic and less tumorigenic. Presumably, the MHC class II molecules presented endogenous tumor specific peptides to $CD4^+$ cells which were needed for tumor rejection and to help activation of $CD8^+$ CTL. For more detailed reviews on MHC molecule expression in tumor cells see Pardoll (1992) and Ostrand-Rosenberg (1994).

6.2
Expression of T Cell Costimulatory Molecules by Tumor Cells

Costimulatory molecules are ligand-receptor pairs found on T cells and antigen presenting cells (APC). One well characterized type of ligand-receptor molecules is the B7 family comprising B7.1 (CD80) and B7.2 (CD86) present on APC. Their counter-receptors are CD28 and CTLA-4 on T cells. Stimulation of T cells by ligation of CD28 leads to activation which includes the production of IL-2 and other cytokines. Several mouse tumor cells were subjected to B7.1 gene transfer. The idea behind this approach is to make tumor cells more potent APC which present tumor antigens to T cells in association with MHC molecules, and in the presence of costimulatory molecules such as B7, this interaction leads to the activation of T cells and the proliferation and expansion of antitumor T cells. The majority of tumor cells do not express B7, and in the absence of B7 the interaction of tumor cells and naive immunocompetent T cells, instead of activating T cells, may lead to T cell anergy, tolerance or apoptosis (Linsley and Ledbetter 1993; Chen et al. 1992; June et al. 1994). Indeed, several B7.1 gene transfected tumors were rejected in syngeneic animals and induced tumor immunity (Chen et al. 1993; Townsend and Allison 1993; Baskar et al. 1993). Depending on whether the tumor cells expressed MHC class I or class II molecules, this effect was mediated by $CD8^+$ or $CD4^+$ T cells. B7.1 expressed by tumor cells seems to act on T cells at the induction and effector phase (Ramarathinam et al. 1994). Consistently, we found that coinjection of $B7.1^+$ and $B7.1^-$ J558L cells could not prevent the growth of $B7.1^-$ tumors unlike what has been observed with cytokine gene transfected cells (unpublished observation). In addition, it was shown that B7.1 expression increased immunogenicity only of such tumors which per se were to some degree immunogenic and expressed MHC class I molecules (Chen et al. 1994). As with cytokines, the level of B7.1 expression seems to be critical. B16-F10 tumor cells were only rejected when they expressed high but not low amounts of B7.1 (Wu al 1995). The expression of costimulatory molecules ICAM-1 and CD2 in B16-F10 cells did not increase the tumor immunogenicity (Dranoff et al. 1993). In order to analyze the reason for the failure of B7.1 to enhance immunogenicity in some tumor models, the B7.1 gene was transfected into $ICAM-1^+$ and $ICAM-1^-$ tumor cell lines which showed that the presence of both molecules was necessary to induce an efficient tumor specific immune response (Cavallo et al. 1995).

7
Tumor Immunity Induced by Genetically Engineered Tumor Cells

7.1
Single-Gene Modification of Tumor Cells

For the clinical perspective the decisive questions are whether genetically engineered tumor cells have vaccine character, how strong the vaccine effect is, how reproducible it is in different tumor models, which genetic modification provides the most effective vaccine and which are the immunological mechanisms underlying tumor immunity. Gene modified tumor cells as vaccine have been tested either by immunization of mice with such cells and subsequent (between 1--4 weeks later) challenge with a tumorigenic dose of the parental tumor cells at a distant site or, in a more rigorous experimental setting, mice first received the parental tumor cells and shortly thereafter were treated with the gene modified cells. Criteria for successful vaccination were rejection of the parental tumor, subcutaneously and in one model orthotopically implanted (Connor et al. 1993), or as shown by Porgador et al. (1992, 1993a,b) suppression of lung metastasis in mice whose primary tumor was removed after having spread. The immunological mechanism, namely the induction of systemic tumor immunity resulting from the local reaction against the gene modified cells, seem to be similar in the prophylactic and therapeutic experiments, even though therapeutic effects were the more difficult to demonstrate the longer the tumor had grown in the mice. For evaluation of the meaning of vaccine experiments it is important to distinguish whether the vaccine effect results from the transfected cytokine or solely from exposure of the mice to the tumor cells. Different tumor cell lines used for vaccine experiments can considerably vary in inherent immunogenicity and immunization with irradiated parental tumor cells occasionally may be sufficient to protect the mice against a challenge with a tumorigenic dose of the same cells (Dranoff et al. 1993).

Do Cytokine Gene-Modified Cells Act as a Vaccine? For several cytokines it has been shown that the respectively transfected cells used for immunization protected more mice from a tumorigenic parental tumor cell challenge in comparison to irradiated or mitomycin C treated parental tumor cells. This has been shown for IL-2, IL-4, IL-6, IL-7, TNF, IFNγ and GM-CSF (Cavallo et al. 1992; Porgador et al. 1992; Hock et al. 1993b; Dranoff et al. 1993; Porgador et al. 1993a,b; Connor et al. 1993; Pericle et al. 1994). The generated immunity was usually tumor specific. It seems as if the transfected cytokine actively contributed to the induction of tumor immunity. However, in one tumor model repeated immunization with the irradiated parental cells could completely compensate for a single injection of cytokine (IL-2) secreting cells (Cavallo et al. 1993).

How Strong Is the Vaccine Effect? This question can be addressed by either increasing the challenge dose or, in the therapeutical setting, asking how long after injection of the parental tumor cells the vaccine cells are still active. We have shown that by increasing the challenge dose the vaccine effect of several cytokine (IL-2, IL-4, IL-7, TNF, IFNγ) producing J558L cells became less obvious (Hock et al. 1993b). When the vaccine effect of cytokine gene transfected cells was compared in a prophylactic and a therapeutic setting, it uniformly showed that the vaccine potency dropped in the latter case, even though vaccination was started within the first 10 days after tumor cell challenge (Golumbek et al. 1991; Dranoff et al. 1993; Connor et al. 1993; Cavallo et al. 1993). The reason for that is not known but most likely is related to tumor burden, tumor induced immune suppression, time needed for tumor immunity, (e.g., CTL activity) to develop, or any combination of these factors. Certainly, a therapeutical model is impeded by the extremely rapid growth of most mouse tumors which does not reflect the situation in most cancer patients. If, however, the decrease or loss of vaccine effect in the therapeutic compared to the prophylactic model results from the difference in inducing an antitumor response in naive animals and breaking tolerance (which seems to be dependent on the presence of the tumor; Schreiber 1993) in tumor bearing animals, then, further treatment modalities which abolish tumor induced immune suppression or reduce tumor load have to precede vaccination. Adoptive immunotherapy with tumor specific T cells of mice with existing tumors was only effective in combination with cyclophosphamide (North 1982; Cheever et al. 1986). Analogous experiments with gene modified tumor cell vaccines have not been reported. In any case, minimal residual disease will be the final target for immunotherapy with vaccines, presumably as an addition to standard therapy.

How Reproducible Is the Vaccine Effect of Cytokine Gene-Modified Tumor Cells? Unfortunately, it is poorly reproducible when the same cytokine is analyzed in different tumor models. This may not be surprising if the above mentioned phenotypical differences of individual tumors are regarded. Some examples for tumor cells transfected to produce either of the cytokines IL-2, IL-4, IL-6, IFNγ or GM-CSF and used either successfully or unsuccessfully in vaccination experiments are shown in Table 7. At the present time, for all cytokines currently employed in clinical trials negative vaccination results have also been reported. There is no reason to assume that similar discrepancies will not also be found in cancer patients. It is clearly of great importance to elucidate the cellular and molecular mechanisms which are responsible for these discrepancies.

By Which Mechanisms Do Gene-Modified Vaccines Induce Tumor Immunity? Whenever analyzed tumor immunity was dependent on T cells. Two models have been proposed by which T cells are induced. In the first, the tumor cells directly present respective antigens to T cells, which are activated in the presence of tumor cell derived T cell stimulatory gene products. In support of this assumption are experiments with IL-2 or B7.1 gene transfected MHC class I^+ class II^- tumors, which showed that tumor immunity was exclusively dependent on $CD8^+$ T cells (Fearon et al. 1990; Chen et al. 1992). However, again no conclusive results have been obtained

Table 7. Poor reproducibility of vaccine effect with gene modified tumor cells

Cytokine gene	Vaccine effect		No vaccine effect	
IL-2	TS/A	(Cavallo et al. 1992)	MCA-102	(Karp et al. 1993)
	MBT-2	(Connor et al. 1993)	B16-F10	(Dranoff et al. 1993)
IL-4	Renca	(Golumbek et al. 1991)	J558L	(Hock et al. 1993b)
	TS/A	(Pericle et al. 1994)	B16-F10	(Dranoff et al. 1993)
IL-6	3LL	(Porgador et al. 1992)	B16-F10	(Dranoff et al. 1993)
			J558L	(unpublished)
IFNγ	3LL	(Porgador et al. 1993a)	J558L	(Hock et al. 1993b)
GM-CSF	B16-F10	(Dranoff et al. 1993)	TS/A	(Allione et al. 1994)
	MBT-2	(Saito et al. 1994)	J558L	(Qin et al. 1996)

IL, interleukin; IFN, interferon; GM-CSF, granulocyte-macrophage colony-stimulating factor.

in different models. Tumor immunity induced by another IL-2 gene trans-fected MHC class I$^+$ class II$^-$ tumor required both, CD4$^+$ and CD8$^+$ T cells although CD4$^+$ T cells were not needed for rejection of the transfected cells (Cavallo et al. 1992). This could indicate that in addition to a direct recognition of the tumor cells by T cells tumor antigens are taken up by APC, processed and presented to CD4$^+$ T cells which in turn may help CD8$^+$ T cells to sustain immunity. The second model is based on experi-ments with GM-CSF transduced tumor cells (Dranoff et al. 1993). GM-CSF induces differentiation/activation of dendritic cells (Steinmann 1991), and the vaccine effect mediated by GM-CSF producing tumor cells has been explained by indirect antigen presentation through APC, possibly dendritic cells, which are induced by GM-CSF to pick up tumor antigen and, after homing into secondary lymphoid organs, e.g., lymph nodes, present the antigens to T cells via both, MHC class I and class II molecules. Consist-ently, CD4$^+$ and CD8$^+$ T cells were required for rejection of the challenge tumor. Indeed, the GM-CSF mediated vaccine effect requires bone marrow derived cells for presentation of even MHC class I antigens as demon-strated by the use of bone marrow chimeras in a tumor model with a surrogate tumor antigen (Levitsky et al. 1994). Furthermore, GM-CSF in-duced the accumulation of dendritic cells at the vaccine site of respectively transfected tumor cells (Qin et al. 1996). In the majority of cases, however, it is not known whether T cells are activated directly by the tumor cells or indirectly by host APC or both pathways and which consequences these two alternatives have on generation of tumor immunity.

7.2
Comparison of Vaccine Effects of Gene-Modified Tumor Cells with Traditional Tumor Vaccines

As new and more sophisticated methods of applying tumor cell vaccines are being continuously developed, it is important to compare them to tumor cell vaccines which have been widely and with little success used in patients. Previously we demonstrated that J558L tumor cells transfected to produce IL-2, IL-4 IL-7, TNF or IFNγ were not qualitatively different at generating systemic immunity and protecting mice against a wildtype tumor cell challenge when compared to the classical vaccine consisting of tumor cells admixed with the adjuvant C. parvum (Hock et al. 1993b; Cayeux et al. 1996). These results were obtained when viable cells either producing the transfected cytokine or mixed with an adjuvant were used for immunization and only those mice which rejected the vaccine cells were evaluated. In essence, similar results were obtained in another tumor model (Cayeux et al. 1995) and Allione et al. (1994) did not find a dramatic difference in immunogenicity between cytokine engineered tumor cells and a tumor cell/adjuvant mixture. These results may not be surprising considering that evolution has developed the appropriate immune response against pathogens which are commonly used as adjuvants. It is not clear, however, whether the best immune response directed against the adjuvants is also the best one directed against tumor cells. Nevertheless, one advantage of using cytokines instead of adjuvants which can be quite different in terms of efficacy from batch to batch (unpublished observation) is standardization of the vaccine. Additionally, the use of cytokines or other immunostimulatory molecules in terms of immunological mechanisms is more clearly defined and by deciphering the type of immune response induced by the various genetic manipulations conditions will eventually be defined which best allow tumor cells to be destroyed. The conclusion is that work on developing potent tumor vaccines has to be continued.

7.3
Improvement of Gene-Modified Tumor Cell Vaccines
by Combinatorial Effects

As we have shown, single gene modification of tumor cells mostly with cytokines or with costimulatory molecules led to improved protection against challenge with wildtype tumors in some models, but in other models modification with one gene was not sufficient to induce an either efficient or reproducible antitumor response. Therefore, various combinatorial strategies have been employed. We have previously found that injection of any possible combination of tumor cells producing either of the cytokines IL-2, IL-4, IL-7, TNF or IFNγ did not augment tumor immunity in comparison to that obtained by one cytokine alone (Hock et al. 1993b). Similarily, coexpression of IL-2 with any of nine different genes encoding immunostimulatory activity did not reveal any synergistic effect (Dranoff et al. 1993).

Recently, we have analyzed the vaccine efficacy of tumor cells cotransfected with genes for B7.1 and cytokines, e.g., IL-4, IL-7. These experiments were based on previous observations that expression of one cytokine in tumor cells most often did not lead to complete tumor elimination when large groups of mice were observed for a prolonged period, that rejection of the transfected tumors did not always correlate with abundance of T cell infiltration even though T cells were needed for rejection or with vaccine potency (Hock et al. 1993a,b). Therefore, the assumption was that T cells in cytokine (IL-2, IL-4, IL-7, TNF, IFNγ) transfected tumors were not appropriately activated. Likewise, B7.1 expression does not reliably cause tumor rejection (Chen et al. 1994). We found that in two tumor models (J558L and TS/A) coexpression of IL-4/B7.1 and IL-7/B7.1, respectively, had an impact on immunogenicity (Cayeux et al. 1995, 1996). In either model expression of cytokine or B7.1 did not ensure reliable rejection of the transfected cells, whereas mice injected with cytokine/B7.1 coexpressing cells always rejected the tumor cells. This effect was mediated by T cells. Moreover, as vaccines TS/A-IL7/B7.1 or J558-IL4/B7.1 cells were more effective than adjuvant/tumor cell mixture or single gene-transfected cells. For IL-7 and B7.1 cotransfected cells, this observation is compatible with in vitro data showing a synergistic effect of IL-7 and anti-CD28 antibodies on T cell proliferation, cytokine secretion and IL-2 receptor expression (Costello et al. 1993). Interestingly, the cytokines induced in this in vitro system included IL-4, TNF and GM-CSF which all have been implicated to either enhance immunogenicity of the respective gene transfected tumor cells or to augment tumor antigen presentation by professional antigen presenting cells such as dendritic cells. When the tumor infiltrating cells were analyzed in single or double gene transfected tumors, we found that in B7.1$^+$ tumors a high percentage of T cells was CD28$^+$ but few T cells were CD25$^+$. In IL-7 tumors CD28$^+$ T cells were virtually absent and a high proportion was CD25$^+$. Both activation markers were found on the majority of T cells in IL-7/B7.1 cotransfected tumors. Although the precise mechanism of the combinatorial use of IL-7 and B7.1 is not yet known, one may speculate that IL-7 and B7.1 act on tumor infiltrating T cells by inducing a cytokine environment which contributes to the efficient host immune response.

An alternative strategy to enhance tumor immunogenicity is the genetic modulation of the tumor cell phenotype, which augments the vaccine effect induced by a transfected cytokine. When we tried to confirm the vaccine effect seen with several GM-CSF gene transfected tumor cells (Dranoff et al. 1993; Saito et al. 1994) in the J558L tumor model, we completely failed (Qin et al. 1996). This discrepancy correlated with the fact that J558L but not the other tumor cells spontaneously secreted IL-10. Because the GM-CSF mediated vaccine effect operates by the involvement of host APC (Huang et al. 1994) and IL-10 has profound downmodulating effects on APC, we have abrogated IL-10 expression in GM-CSF producing J558L cells. Indeed, the resulting cells acted as vaccine and, furthermore, we showed that IL-10 inhibited the GM-CSF induced accumulation of dendritic cells at the vaccine site.

As a third example, we showed that coexpression of GM-CSF and TNF in J558L cells synergistically induced tumor immunity (manuscript in preparation). Together, based on the knowledge of the tumor cell phenotype on one hand and the mechanisms of how to induce most potently and specifically an (T cell) immune response in the future more sophisticated and precise strategies to construct potent tumor vaccines can be elaborated.

7.4
Viable Versus Nonproliferating Tumor Cell Vaccines

Present clinical protocols use irradiated gene modified (100 Gy) tumor cell vaccines. We compared the efficiency of irradiated versus live IL7/B7.1 or IL4/B7.1 tumor cell vaccines to protect against tumor cell challenge and found that live cells generated a much better systemic tumor immunity than irradiated cells (Cayeux et al. 1995, 1996; Fig. 4). This was not caused by the loss of gene expression, both cytokine and B7.1 were still expressed in vitro 10 days after irradiation. Similar results were previously obtained with tumor cell/adjuvant as vaccine (Hock et al. 1993b) or with cytokine or B7 transfected tumor cells in other models (Allione et al. 1994; Townsend et al. 1994). No clear explanation for this finding is yet available but this loss of vaccine efficacy could be due to release of inhibitory immunosuppressive factors by tumor cells after irradiation or interference with antigen processing or presentation by irradiation. In contrast, GM-CSF transduced

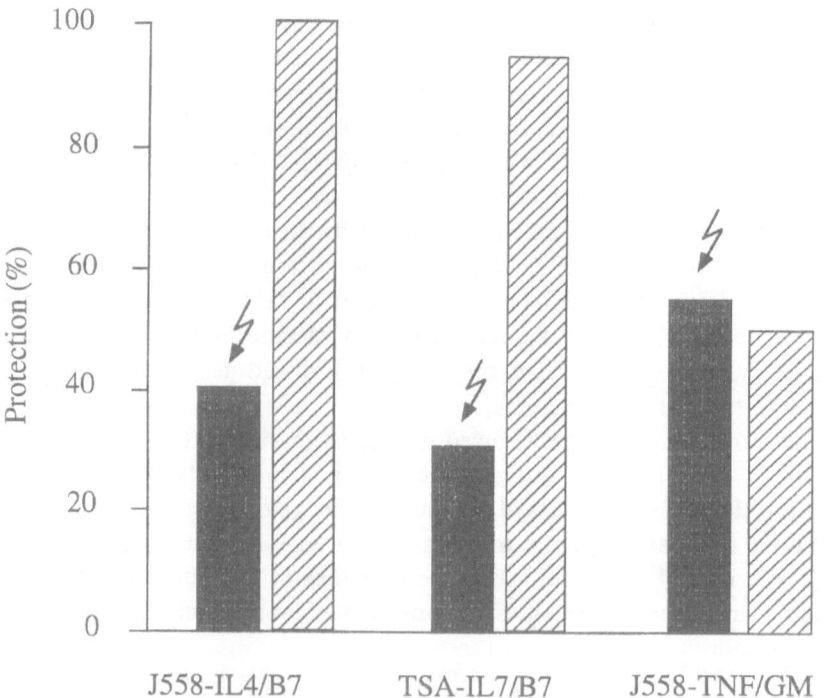

Fig. 4. Influence of irradiation on vaccination efficacy. The same numbers of irradiated (*filled column*) or viable (*hatched column*) vaccine cells transfected with genes, as indicated, were used to immunize syngeneic mice. Two weeks later, mice were contralaterally challenged with parental cells. Rejection of the challenge tumor was recorded as percentage protection. For details, see Hock et al. (1993b), Cayeux et al. (1995, 1996), and manuscript in preparation

B16-F10 (Dranoff et al. 1993) or GM-CSF/TNF transduced J558L cells (manuscript in preparation) were equally effective regardless whether the vaccine cells were irradiated or not (Fig. 4). This comparison was possible because GM-CSF producing B16-F10 cells were rejected when additionally IL-2 was expressed by the tumor cells (Dranoff et al., 1993) and GM-CSF/TNF producing J558L cells did not grow in syngeneic mice because of TNF production (Blankenstein et al. 1991a). Notably, the results with IL-4/B7.1 and GM-CSF/TNF transfected cells were obtained in the same tumor model, and equal numbers of cells were used for immunization. This may indicate that the different vaccine efficacy between irradiated and viable cells is not caused by proliferation and increase in antigen concentration before the cells are rejected, even though one cannot exclude that higher amounts of antigen can compensate for the reduction of immunogenicity by irradiation. These results could rather be explained by the above mentioned way of antigen presentation to T cells, directly by the IL4/B7.1 transfected tumor or indirectly in case of GM-CSF.

8
Current Clinical Trials and Anticipated Problems

It is clear that the experimental models which have been discussed so far are not only of academic interest but also for clinical application. Despite the continuous improvements of therapeutic strategies, experimental tumor vaccine models have revealed some persisting problems. The two most important of them are reproducibility in different tumor models and decay of the immune response in tumor bearing hosts. However, from the perspective of many cancer patients a rapid translation of experimental vaccine strategies into clinical application would be welcome considering the limited curative potential of chemo/radiation therapy. This argument alone is sufficient to justify clinical trials, but it does not address the question of scientific value. A number of clinical trials employing cytokine gene modified tumor cells as vaccines have begun. Cytokines used are IL-2, IL-4, IL-7, TNF, IFNγ and GM-CSF (Tepper and Mulé 1994). Instead of listing the multitude of possibilities which can be responsible for interference with successful immunotherapy and which have exhaustively been reviewed by Schreiber (1993), it seems important to ask which of the anticipated problems of the clinical trials can also be found in mice and which are specific to humans.

Problems which potentially can interfere with tumor vaccination and are common to humans and mice are:

1. *Mechanisms of tumor escape from immunological destruction* (loss of ability of tumor cells to present suitable antigens and inability of host immune cells to become activated or reach the tumor site). The various tumor or host related escape mechanisms, e.g., downregulation of MHC expression, production of immunosuppressive factors, immunoselection of tumor cell variants etc. (Schreiber 1993) can be elucidated in well-controlled mouse models and methods to inhibit such escape will guide to more specific and reproducible immunotherapy in the clinic.

2. *Decay of the immune response in tumor bearing hosts.* In many experiments (see above) it was shown that gene modified tumor cells were far less effective for the treatment of tumor bearing animals than for the prevention of immunized mice to develop tumor after challenge. It seems likely that the same problem will also be found in cancer patients. The conclusion for that is that it will be important to work out the responsible mechanisms and to resolve the question how tumor induced immunosuppression can be counteracted. Furthermore, it is suggested that immunotherapy with tumor vaccines will most likely be effective in patients with low tumor load, e.g., minimal residual disease.

3. *Difference in tumor cell phenotype which influence the immune response and impede reproducibility.* As it has been shown in Table 7, immunization with different tumor cell lines transfected with the same cytokine can yield quite different results. The phenotype of the respective tumors may largely contribute to these discrepancies. To work out the responsible mechanisms will more clearly define the type of tumors susceptible to immunotherapy in the clinic.

Problems of immunotherapy which may be specific to humans but not to mice:

1. *Outbred situation.* Vaccination experiments have always been done in inbred mice with syngeneic tumors. No data do exist whether any of the above mentioned different results are due to the genetic background of the mice. In humans, genetic variations of the patients have to be taken into account.

2. *Age of the host.* To our knowledge, experimental tumor models have always been carried out in young animals, whereas most cancers occur

in aged patients. It is currently not known, whether and in which way this makes a difference.

3. *Time needed to evaluate therapeutic efficacy.* Essentially, what is important to demonstrate in patients, is prolongation of survival attributed to immunotherapy. Considering that those patients who are the most promising candidates for immunotherapy are those with minimal residual disease, it becomes apparent that relapse may be expected to occur relatively late. Consequently, conclusions to be drawn with regard to therapeutic efficacy may take a long time.

4. *Criteria to rapidly evaluate therapeutic efficacy.* Because success of current gene therapy approaches in the clinic is not clear at all at the moment and new strategies are constantly been developed, it is important to find reliable parameters which allow rapid and reproducible analysis of specific effects resulting from immunization. The search for tumor specific T cells (or T cell receptors) is one option, provided they can be clearly defined.

5. *Practicability.* Current clinical trials employ ex vivo gene transfer, most in an autologous and some in an allogeneic setting. An autologous vaccine requires a tumor cell culture to be established from each individual patient. This is rather labor- and cost-intensive and may be possible for some tumors but not for others. The major disadvantage, however, is that the vaccines differ as much from each other as different tumors within one histologic type can differ from each other and that due to lack of time the vaccine cells cannot carefully be analyzed. Therefore, standardization of the vaccine cells is not possible and a correlation between a potential antitumor response and phenotype of the vaccine is difficult. In terms of feasibility and standardization, an allogeneic vaccine seems more convenient. It requires the knowledge of (specific or associated) tumor antigen(s) present on the vaccine cells and shared with those on the patient's tumor to be treated, the knowledge of the presenting MHC restriction element and methods to detect the antigens. Therefore, a partial match of HLA haplotype between patient and vaccine cells should exist. It is currently not known whether the expected allogeneic immune response interferes with induction of tumor specific CTL.

6. *Schedule of treatment.* It is largely unknown how many vaccine cells have to be injected and what the duration of treatment should be in order to be effective.

9
Conclusion

A new era of immunotherapy of cancer has begun, often referred to as gene therapy. The general concepts have not changed, but the available methods and reagents and the understanding of how immune responses are induced and regulated. Of importance are the following: (a) a number of gene transfer techniques, (b) the discovery of an array of cloned and functionally characterised cytokines, (c) the knowledge of how (tumor) cells can present peptides of potentially any cellular protein to T cells by MHC molecules, (d) the basic principles of activation of T cells which are believed to play a central role in immunotherapy and (e) molecular mechanisms of malignant transformation. A multitude of experimental strategies for cancer immunotherapy are employed which are based on solid ground as was never before the case. Most experimental work has been done in order to construct tumor cell vaccines by genetic engineering with cytokines and/or other immunostimulatory proteins such as B7. In mouse tumor models it has been shown that gene modified tumor cell vaccines can be more effective than traditional tumor vaccines which they are going to replace in the clinic. However, it should not be withholded that experimental tumor models have revealed some chronic problems in tumor immunology, that is the individuality of each tumor which makes general conclusion difficult and the rapid modulation of the host's immune system by the tumor. If the flood of new and important information during the last 5 years is considered, one can optimistically assume that these problems can be resolved, but the clue to successful cancer immunotherapy in most cases has to come from well-defined experimental models.

Acknowledgments. We wish to thank Günther Richter and Gerald Willimsky for critical reading of the manuscript. This work was supported by grants from the Deutsche Krebshilfe, Mildred Scheel Stiftung e.V. and the BMBF.

References

Allione A, Consalvo M, Nanni P et al (1994) Immunizing and curative potential of replicating and nonreplicating murine mammary adenocarcinoma cells engineered with interleukin (IL)-2, IL-4, IL-6, IL-7, IL- 10, tumor necrosis factor α, granulocyte-macrophage colony-stimulating factor, and γ-interferon gene or admixed with conventional adjuvants. Cancer Res 54:6022–6026

Anderson WF (1984) Prospects for human gene therapy. Science 226:401–409

Anderson WF (1992) Human gene therapy. Science 256:808–813

Aoki T, Tashiro K, Miyatake SIY et al (1992) Expression of murine interleukin 7 in a murine glioma cell line results in reduced tumorigenicity in vivo. Proc Natl Acad Sci USA 89:3850–3854

Asher AL, Muli AL, Kasid A et al (1991) Murine tumor cells transduced with the gene for tumor necrosis factor-α. J Immunol 146:3227–3234

Barth RJ, Muli JJ, Spiess PJ et al (1991) Interferon γ and tumor necrosis factor have a role in tumor regression mediated by murine CD8+ tumor-infiltrating lymphocytes. J Exp Med 173:647–658

Baskar S, Ostrand-Rosenberg S, Nabavi N et al (1993) Constitutive expression of B7 restores immunogenicity of tumor cells expressing truncated MHC class II molecules. Proc Natl Acad Sci USA 90:5687–5690

Bernards R (1987) Suppression of MHC gene expression in cancer cells. Trends Genet 3:298–301

Blaese RM, Ishii-Morita H, Mullen C et al (1994) In situ delivery of suicide genes for cancer treatment. Eur J Cancer 30A:1190–1193

Blankenstein T (1994) Increasing tumour immunogenicity by genetic modification. Eur J Cancer 30A:1182–1187

Blankenstein T, Qin Z, Überla K et al (1991a) Tumor suppression after tumor cell targeted tumor necrosis factor alpha gene transfer. J Exp Med 173:1047–1052

Blankenstein T, Rowley DA, and Schreiber H (1991b) Cytokines and cancer: experimental systems. Curr Opin Immunol 3:694–698

Boon T, Cerottini JC, Van den Eynde B et al (1994) Tumor antigens recognized by T lymphocytes. Annu Rev Immunol 12:337–365

Bottazzi B, Walter S, Govoni D et al (1992) Monocyte chemotactic cytokine gene transfer modulation, macrophage infiltration, growth, and susceptibility to IL-2 therapy of a murine melanoma. J Immunol 148:1280–1285

Brenner MK, Rill DR, Heslop et al (1994) Gene marking after bone marrow transplantation. Eur J Cancer 30A:1171–1176

Bubenik J, Voitenok NN, Kieler J et al (1988) Local administration of cells containing an inserted IL-2 gene and producing IL-2 inhibits growth of human tumors in nu/nu mice. Immunol Lett 19:279–282

Bubenik J, Simova J, Jandlova T (1990) Immunotherapy of cancer using local administration of lymphoid cells transformed by IL-2 cDNA and constitutively producing IL-2. Immunol Lett 23:287–292

Cavallo F, Giovarelli M, Gulino A et al (1992) Role of neutrophils and CD4+ T lymphocytes in the primary and memory response to nonimmunogenic murine mammary adenocarcinoma made immunogenic by IL-2 gene transfection. J Immunol 149:3627–3635

Cavallo F, Di Pierro F, Giovarelli M et al (1993) Protective and curative potential of vaccination with interleukin-2-gene- transfected cells from a spontaneous mouse mammary adenocarcinoma. Cancer Res 53:5067–5070

Cavallo F, Martin-Fontecha A, Bellone M et al (1995) Co-expression of B7-1 and ICAM-1 on tumors is required for rejection and the establishment of a memory response. Eur J Immunol 25: 1154–1162

Cayeux S, Beck C, Aicher A et al (1995) Tumor cells cotransfected with interleukin 7 and B7.1 genes induce CD25 and CD28 on tumor infiltrating lymphocytes and are strong vaccines. Eur J Immunol 25:2325–2331

Cayeux S, Beck C, Dörken B et al (1996) Coexpression of interleukin 4 and B7.1 in murine tumor cells leads to improved tumor rejection and vaccine effect compared to single gene transfectant and a classical adjuvant. Hum Gene Therapy 7:525–529

Chen L, Ashe S, Brady WA et al (1992) Costimulation of antitumor immunity by the B7 counterreceptor for the T lymphocyte molecules CD28 and CTLA-4. Cell 71:1093–1102

Chen L, Linsley PS, Hellström KE (1993) Costimulation of T cells for tumor immunity. Immunol Today 14: 482–486

Chen L, McGowan P, Ashe S et al (1994) Tumor immunogenicity determines the effect of B7 costimulation on T cell–mediated tumor immunity. J Exp Med 179:523–532

Cheever MA, Thompson DB, Klarnet JP et al (1986) Antigen-driven long term-cultured T cells proliferate in vivo, destribute widely, mediate specific tumor therapy, and persist long-term as functional memory T cells. J Exp Med 163:1100–1112

Colombo MP, Ferrari G, Stoppacciaro A et al (1991) Granulocyte colony-stimulating factor gene transfer suppresses tumorigenicity of a murine adenocarcinoma in vivo. J Exp Med 173:889–897

Colombo MP, Modesti A, Parmiani G et al (1992) Local cytokine availability elicits tumor rejection and systemic immunity through granulocyte-T-lymphocyte crosstalk. Can Res 52:4853–4857

Connor J, Bannerji R, Saito S et al (1993) Regression of bladder tumors in mice treated with interleukin 2 gene-modified tumor cells. J Exp Med 177:1127–1134

Costello R, Brailly H, Mallet F et al (1993) Interleukin-7 is a potent co-stimulus of the adhesion pathway involving CD2 and CD28 molecules. Immunology 80:451–457

Counoyer D, Caskey CT (1993) Gene therapy of the immune system. Annu Rev Immunol 11:297–329

Dalgleish AG (1994) Cancer vaccines. Eur J Cancer 30A:1029–1035

Dorsch M, Hock H, Kunzendorf U et al (1993) Macrophage colony-stimulating factor gene transfer into tumor cells induces macrophage infiltration but not tumor suppression. Eur J Immunol 23:186–190

Douvdevani A, Huliehel M, Zvller M et al (1992) Reduced tumorigenicity of fibrosarcomas which constitutively generate IL-1α either spontaneously or following IL-1α gene transfer. Int J Cancer 52:1–9

Dranoff G, Jaffee E, Lazenby A et al (1993) Vaccination with irradiated tumor cells engineered to secrete murine granulocyte-macrophage colony-stimulating factor stimulates potent, specific, and long-lasting anti-tumor immunity. Proc Natl Acad Sci USA 90:3539–3543

Esumi N, Hunt B, Itaya T et al (1991) Reduced tumorigenicity of murine tumor cells secreting interferon-γ is due to nonspecific host responses and is unrelated to class I major histocompatibility complex expression. Cancer Res 51:1185–1189

Fearon ER, Pardoll DM, Itaya T et al (1990) Interleukin 2 production by tumor cells bypasses T helper function in the generation of an antitumor response. Cell 60:397–403

Fefer A, Truitt RL, Sullivan KM (1991) Adoptive cellular therapy: graft-versus-tumor responses after bone-marrow transplantation. In: DeVita VT, Helman S, Rosenberg SA (eds) Biologic therapy of cancer, principles and pratice. Lippincott, New York, pp 237–246

Ferrantini M, Proietti E, Santodonato GL et al (1993) α₁-Interferon gene transfer into metastatic friend leukemia cells abrogated tumorigenicity in immunocompetent mice: antitumor therapy by means of interferon-producing cells. Cancer Res 53:1107–1112

Ferrantini M, Giovarelli M, Modesti A et al (1994) IFN-α1 gene expression into a metastatic murine adenocarcinoma (TS/A) results in CD8$^+$ T cell-mediated tumor rejection and development of antitumor immunity. Comparative studies with IFN-γ-producing TS/A cells. J Immunol 153(10):4604–4615

Flamand V, Sornasse T, Thielemans K et al (1994) Murine dendritic cells pulsed in vitro with tumor antigen induce tumor resistance in vivo. Eur J Immunol 24:605–610

Friedmann T (1989) Progress toward human gene therapy. Science 244:1275–1281

Gansbacher B, Zier K, Daniels B et al (1990a) Interleukin 2 gene transfer into tumor cells abrogates tumorigenicity and induces protective immunity. J Exp Med 172:1217–1224

Gansbacher B, Bannerji R, Daniels B et al (1990b) Retroviral vector-mediated interferon-γ gene transfer into tumor cells generates potent and long lasting antitumor immunity. Cancer Res 50:7820–7825

Grard CM, Bruyns C, Delvaux A et al (1996) Loss of tumorigenicity and increased immunogenicity induced by interleukin-10 gene transfer in B16 melanoma cells. Human Gene Ther 7:23–31

Giovarelli M, Musiani P, Modesti A et al (1995) Local release of IL-10 by transfected mouse mammary adenocarcinoma cells does not suppress but enhances antitumor reaction and elicits a strong cytotoxic lymphocyte and antibody-dependent immune memory. J Immunol 155:3112–3123

Golumbek PT, Lazenby AJ, Levitsky HI et al (1991) Treatment of established renal cancer by tumor cells engineered to secrete interleukin 4. Science 254:713–716

Greenberg PD (1991) Adoptive T cell therapy of tumors: mechanisms operative in the recognition and elimination of tumor cells. Adv Immunol 49: 281–355

Herr HW (1991) Instillation therapy for bladder cancer. In: DeVita VT, Helman S, Rosenberg SA (eds) Biologic therapy of cancer, principles and practice. Lippincott, New York, pp 643–650

Hock H, Dorsch M, Diamantstein T et al (1991) Interleukin 7 induces CD4$^+$ T cell-dependent tumor rejection. J Exp Med 174:1291–1298

Hock H, Dorsch M, Kunzendorf U et al (1993a) Mechanisms of rejection induced by tumor cell targeted gene transfer of interleukin-2, interleukin-4, interleukin-7, tumor necrosis factor or interferon-gamma. Proc Natl Acad Sci USA 90:2774–2778

Hock H, Dorsch M, Kunzendorf et al (1993b) Vaccinations with tumor cells genetically engineered to produce different cytokines: effectivity not superior to a classical adjuvant. Cancer Res 53:714–716

Hoover HC, Hanna MG (1991) Immunotherapy by active specific immunization: clinical applications. Colon cancer. In: DeVita VT, Helman S, Rosenberg SA (eds) Biologic therapy of cancer, prinicples and practice. Lippincott, New York, pp 670–681

Huang AYC, Golumbek P, Ahmadzadeh M et al (1994) Role of bone marrow-derived cells in presenting MHC class I-restricted tumor antigens. Science 264:961–965

Hui K, Grosveld F, Festensein H et al (1984) Rejection of transplantable AKR leukemic cells following MHC DNA-mediated cell transformation. Nature 311:750–752

Hwu P, Shafer GE, Treisman J et al (1993) Lysis of ovarian cancer cells by human lymphocytes redirected with a chimeric gene composed of an antibody variable region and the Fc receptor γ chain. J Exp Med 178:361- 366

Itaya T, Yamagiwa S, Okada F et al (1987) Xenogenization of a mouse lung carcinoma (3LL) by transfection with an allogeneic class II major histocompatibility complex gene (H-2L^d). Cancer Res 47:3136–3140

James R, Edwards S, Hui K et al (1991) The effect of class II gene transfection on the tumorigenicity of the H-2 K negative mouse leukemia cell line K36.16. Immunology 72:213–218

June CH, Bluestone JA, Nadler LM et al (1994) The B7 and CD28 receptor families. Immunol Today 15:321–331

Kantoff PW, Freeman M, Anderson WF et al (1988) Prospects for gene therapy for immunodeficiency diseases. Annu Rev Immunol 6:581–594

Karp SE, Farber A, Salo JC et al (1993) Cytokine secretion by genetically modified nonimmunogenic murine fibrosarcoma. Tumor inhibition by IL2 but not tumor necrosis factor. J Immunol 150:896–908

Kärre K, Ljunggren H, Piontek G et al (1986) Selective rejection of H-2 deficient lymphoma variants suggests alternative immune defense strategy. Nature 319:675–686

Katsanis E, Orchard PJ, Bausero MA et al (1994) Interleukin-2 gene transfer into murine neuroblastoma decreases tumorigenicity and enhances systemic immunity causing regression of preestablished retroperitoneal tumors. J Immunother 15:81–90

Khazaie K, Prifti S, Beckhove P et al (1994) Persistence of dormant tumor-cells in the bone marrow of tumor-cell-vaccinated mice correlates with long term immunological protection. Proc Natl Acad Sci USA 91:7430–7434

Krüger-Krasagakes S, Li W, Richter G et al (1993) Eosinophils infiltrating interleukin 5 gene transfected tumors do not suppress tumor growth. Eur J Immunol 23:992–995

Lang RA, Burgess AW (1990) Autocrine growth factors and tumourigenic transformation. Immunol Today 7:244–249

Leach DR, Callahan GN (1995) Fibrosarcoma cells expressing allogeneic MHC class II antigens induce protective antitumor immunity. J Immunol 154:738–743

Lebel-Binay S, Laguerre B, Quintin-Colonna F et al (1995) Experimental gene therapy of cancer using tumor cells engineered to secrete interleukin-13. Eur J Immunol 25:2340–2348

Levitsky HI, Lazenby A, Hayashi RJ et al (1994) In vivo priming of two distinct antitumor effector populations: the role of MHC class I expression. J Exp Med 179: 1215–1224

Levy R, Miller RA (1991) B-cell lymphomas. In: DeVita VT, Helman S, Rosenberg SA (eds) Biologic therapy of cancer, principles and practice. Lippincott, New York, p 275

Ley V, Langlade-Demoyen P, Kourilsky P et al (1991) Interleukin 2-dependent activation of tumor-specific cytotoxic T lymphocytes in vivo. Eur J Immunol 21:851–854

Liebermann DA, Hoffmann B, Steinmann RA (1995) Molecular controls of growth arrest and apoptosis: p53-dependent and independent pathways. Oncogene 11:199–210

Li W, Diamantstein T, and Blankenstein T (1990) Lack of tumorigenicity of interleukin 4 autocrine growing cells seems related to the anti-tumor function of interleukin 4. Mol Immunol 27:1331–1337

Linsley PS, Ledbetter JA (1993) The role of the CD28 receptor during T cell response to antigen. Annu Rev Immunol 11:191–212

Lollini PL, Bosco MC, Cavallo F et al (1993) Inhibition of tumor growth and enhancement of metastasis after transfection of the γ-interferon gene. Int J Cancer 55:320–329

Lotze MT, Rosenberg SA (1991) Interleukin-2: clinical applications. In: DeVita VT, Helman S, Rosenberg SA (eds) Biologic therapy of cancer, principles and practice. Lippincott, New York, p 159

Luster AD, Leder P (1993) IP-10, a c-x-c-chemokine, elicits a potent thymus-dependent antitumor response in vivo. J Exp Med 178: 1057–1065

Mantovani A, Bottazzi B, Colotta F et al (1992) The origin and function of tumor-associated macrophages. Immunol Today 13:265–270

Martinotti A, Stoppacciaro A, Vagliani M et al (1995) CD4 T cells inhibit in vivo the CD8-mediated immune response against murine colon carcinoma cells transduced with interleukin-12 genes. Eur J Immunol 25:137–146

McBride WH, Thacker JD, Comora S et al (1992) Genetic modification of a murine fibrosarcoma to produce interleukin 7 stimulates host cell infiltration and tumor immunity. Cancer Res 52:3931–3937

McBride WH, Dougherty DG, Wallis AE et al (1994) Interleukin-3 in gene therapy of cancer. Folia Biologica 40:62–73

Melief CJM (1992) Tumor eradication by adoptive transfer of cytotoxic T lymphocytes. Adv Cancer Res 58:143–175

Miller AD (1992) Human gene therapy comes of age. Nature 357:455–460

Miller AR, McBride WH, Dubinett SM et al. (1993) Transduction of human melanoma cell lines with the human interleukin-7 gene using retroviral-mediated gene transfer: comparison of immunologic properties with interleukin-2. Blood 82/12:3686–3694

Moormeier JA, Golomb HM (1991) Interferons – clinical applications. In: DeVita VT, Helman S, Rosenberg SA (eds) Biologic therapy of cancer, principles and practice. Lippincott, New York, p 275

Morton DL, Foshag LK, Hoon DSB et al (1992) Prolongation of survival in metastatic melanoma after active specific immunotherapy with a new polyvalent melanoma vaccine. Ann Surgery 216:463–482

Mullen CA, Coale MM, Levy AT et al (1992) Fibrosarcoma cells transduced with the IL-6 gene exhibit reduced tumorigenicity, increased immunogenicity, and decreased metastatic potential. Cancer Res 52:6020–6024

Mulligan RC (1993) The basic science of gene therapy. Science 260:926–932

North RJ (1982) Cyclophosphamide-facilitated adoptive immunotherapy of an established tumor depends on elimination of tumor induced suppressor T cells. J Exp Med 55:1063–1074

Oettgen HF, Old LJ (1991) The history of cancer immunotherapy. In De Vita VT, Helman S, Rosenberg SA (eds) Biologic therapy of cancer, principles and practice. Lippincott, New York, pp 87–119

Ohe Y, Podack ER, Olsen KJ et al (1993) Combination effect of vaccination with IL2 and IL4 cDNA transfected cells on the induction of a therapeutic immune response against Lewis lung carcinoma cells. Int J Cancer 53:432–437

Oliff A, Defeo-Jones D, Boyer M et al (1987) Tumors secreting human TNF/cachectin induce cachexia in mice. Cell 50:555- 563

Ostrand-Rosenberg S (1994) Tumor immunotherapy: the tumor cell as an antigen-presenting cell. Curr Opinion Immunol 6:722–727

Ostrand-Rosenberg S, Thakur A, Clements V (1990) Rejection of mouse sarcoma cells after transfection of MHC class II genes. J Immunol 144:4068–4071

Pardoll D (1992) New strategies for active immunotherapy with genetically engineered tumor cells. Curr Opinion Immunol 4:619–623

Pardoll D (1993) Cancer vaccines. Immunol Today 6:310–316

Pardoll DM (1994) A new look for the 1990s. Nature 369:357–358

Pekarek LA, Starr BA, Toledano AY et al (1995) Inhibition of tumor growth by elimination of granulocytes. J Exp Med 181:435–440

Pericle F, Giovarelli M, Colombo MP et al (1994) An efficient Th2-type memory follows CD8$^+$ lymphocyte-driven and eosinophil-mediated rejection of a spontaneous mouse mammary adenocarcinoma engineered to release IL-4. J Immunol 153:5659–5673

Platzer C, Richter G, Überla K et al (1992) Interleukin 4 mediated tumor suppression in nude mice involves interferon-gamma. Eur J Immunol 22:1729–1733

Porgador A, Tzehoval E, Katy VE et al (1992) Interleukin 6 gene transfection into Lewis lung carcinoma tumor cells suppresses the malignant phenotype and confers immunotherapeutic competence against parental metastatic cells. Cancer Res 52:3679–3686

Porgador A, Bannerji R, Watanabe Y et al (1993a) Anti-metastatic vaccination of tumor-bearing mice with two types of γIFN gene inserted tumor cells. J Immunol 150:1458–1470

Porgador A, Bannerji R, Tzehoval E et al (1993b) Anti-metastatic vaccination of tumor-bearing mice with IL-2 gene inserted tumor cells. Int J Cancer 53:471–477

Pulaski BA, McAdam AJ, Hutter EK et al (1993) Interleukin 3 enhances development of tumor-reactive cytotoxic cells by a CD4-dependent mechanism. Cancer Res 53:2112–2117

Qin Z, Krüger-Krasagakes S, Kunzendorf U et al (1993) Expression of tumor necrosis factor by different tumor cell lines results either in tumor suppression or augmented metastasis. J Exp Med 178:355–360

Qin Z, Blankenstein T (1995) Tumor growth inhibition mediated by lymphotoxin: evidence of B lymphocyte involvement in the antitumor response. Cancer Res 55: 4747–4751

Qin Z, van Tits LJH, Buurman WA et al (1995) Human lymphotoxin has at least equal antitumor activity in comparison to human tumor necrosis factor but is less toxic in mice. Blood 85:2779–2785

Qin Z, Noffz G, Mohaupt M et al (1996) Interleukin 10 prevents dendritic cell infiltration and vaccination with granulocyte-macrophage colony-stimulating factor gene modified tumor cells (submitted)

Quinnan GVJ, Kirmani N, Rook AH et al (1982) Cytotoxic T cells in cytomegalovirus infection: HLA-restricted T-lymphocyte and non-T-lymphocyte cytotoxic responses correlate with recovery from cytomegalovirus infection in bone-marrow-transplant recipients. N Engl J Med 307:7–13

Ramarathinam L, Castle M, Wu Y et al (1994) T cell costimulation by B7/BB1 induces CD8 T cell-dependent tumor rejection: an important role of B7/BB1 in the induction, recruitment, and effector function of antitumor T cells. J Exp Med 179:1205–1214

Renner C, Jung W, Sahin U et al (1994) Cure of xenografted human tumors by bispecific monoclonal antibodies and human T cells. Science 264:833–835

Restifo NP, Spiess PJ, Karp SE et al (1992) A nonimmunogenic sarcoma transduced with the cDNA for interferon-γ elicits CD8[+] T cells against the wild-type tumor: correlation with antigen presentation capability. J Exp Med 175:1423–1431

Richter G, Krüger-Krasagakes S, Hein G et al (1993) Interleukin 10 transfected into Chinese hamster ovary prevents tumor growth and macrophage infiltration. Cancer Res 53:4134–4137

Riddel SR, Watanabe KS, Goodrich JM et al (1992) Restoration of viral immunity in immunodeficient humans by the adoptive transfer of T cell clones. Science 257:238–241

Riethmüller G, Schneider-Gädicke E, Schlimok G et al: German Cancer Aid 17-A Study Group (1994) Randomised trial of monoclonal antibody for adjuvant therapy of resected Dukes' C colorectal carcinoma. Lancet 343:1177–1183

Rollins BJ, Sunday ME (1991) Suppression of tumor formation in vivo by expression of the JE gene in malignant cells. Mol Cell Biol 11:3125–3131

Rosenberg SA (1991) Adoptive cellular therapy: clinical applications. In: DeVita VT, Helman S, Rosenberg SA (eds) Biologic therapy of cancer, principles and practice. Lippincott, New York, p 214

Russell SJ, Eccles SA, Flemming CL et al (1991) Decreased tumorigenicity of a transplantable rat sarcoma following transfer and expression of an IL-2 cDNA. Int J Cancer 47:244–251

Saito S, Bannerji R, Gansbacher B et al (1994) Immunotherapy of bladder cancer with cytokine gene-modified tumor vaccines. Cancer Res 54:3516–3520

Schreiber H (1993) Tumor immunology. In: Paul WE (ed) Fundamental immunology, 3rd edn. Raven, New York, pp 1143–1178

Sporn MB, Roberts AB (1985) Autocrine growth factors and cancer. Nature 313:745–747

Steinman RM (1991) The dendritic cell system and its role in immunogenicity. Annu Rev Immunol 9:271–296

Stoppacciaro A, Melani C, Parenza M et al (1993) Regression of an established tumor genetically modified to release granulocyte colony-stimulating factor requires granulocyte-T cell cooperation and T cell-produced interferon-γ. J Exp Med 178:151–161

Suzuki T, Tahara H, Narula S et al (1995) Viral interleukin 10 (IL-10), the human herpes virus 4 cellular IL-10 homologue, induces local anergy to allogeneic and syngeneic tumors. J Exp Med 182:477–486

Tahara H, Zitvogel L, Storkus W-J et al (1995) Effective eradication of established murine tumors with IL-12 gene therapy using a polycistronic retroviral vector. J Immunol 154:6466–6474

Takahashi H, Nakada T, Puisieux I, et al. (1993) Inhibition of human colon cancer growth by antibody-directed human LAK cells in scid mice. Science 259:1460–1463

Tanaka K, Isselbacher K, Khoury G et al (1985) Reversal of oncogenesis by the expression of a major histocompatibility complex class I gene. Science 228: 26–30

Taniguchi K, Petersson M, Hvglund P et al (1987) Interferon-γ induces lung colonization by intravenously inoculated B16 melanoma cells in parallel with enhanced expression of class I major histocompatibility complex antigens. Proc Natl Acad Sci USA 84:3405–3409

Tao MH, Levy R (1993) Idiotype/granulocyte-macrophage colony-stimulating factor fusion protein as a vaccine for B-cell lymphoma. Nature 362:755–758

Teng MN, Park BH, Koeppen HKW et al (1991) Long-term inhibition of tumor growth by tumor necrosis factor in the absence of cachexia or T-cell immunity. Proc Natl Acad Sci USA 88:3535–3539

Tepper RI, Mul JJ (1994) Experimental and clinical studies of cytokine gene-modified tumor cells. Hum Genet Ther 5:153–164

Tepper RI, Pattengale PK, and Leder P (1989) Murine interleukin 4 displays potent anti-tumor activity in vivo. Cell 57:503- 512

Tepper RI, Coffman RL, and Leder P (1992) An eosinophil-dependent mechanism for the anti-tumor effect of interleukin 4. Science 257:548–551

Torre-Amione G, Beauchamp RD, Koeppen H et al (1990) A highly immunogenic tumor transfected with a murine transforming growth factor β1 cDNA escapes immune surveillance. Proc Natl Acad Sci USA 87:1486–1490

Townsend SE, Allison JP (1993) Tumor rejection after direct costimulation of CD8+ T cells by B7-transfected melanoma cells. Science 259:368–370

Townsend SE, Su FW, Atherton JM et al (1994) Specificity and longevity of antitumor immune responses induced by B7-transfected tumors. Cancer Res 54:6477–6483

Tsai S, Gansbacher B, Tait L et al (1993) Induction of antitumor immunity by interleukin 2 gene-transduced mouse mammary tumor cells versus transduced mammary stromal fibroblasts. J Natl Cancer Inst 85:546- 553

Verma IM (1990) Gene therapy. Sci Am 263:68–72

Visseren MJW, Koot M, van der Voort EIH et al (1994) Production of interleukin-2 by EL4 tumor cells induces natural killer cell- and T-cell-mediated immunity. J Immunother 15:119–128

Waldman TA (1991) T-cell leukemia/lymphoma. In: DeVita VT, Helman S, Rosenberg SA (eds) Biologic therapy of cancer, principles and practice. Lippincott, New York, p 523

Wallich R, Bulbuc N, Hammerling G et al (1985) Abrogation of metastatic properties of tumor cells by de novo expression of H-2 K antigens following H-2 gene transfection. Nature 315:301–305

Watanabe Y, Kuribayashi K, Miyatake S et al (1989) Exogenous expression of mouse interferon-γ cDNA in mouse neuroblastoma C1300 cells results in reduced tumorigenicity by augmented anti-tumor immunity. Proc Natl Acad Sci USA 86:9456–9460

Wu LC, Huang AYC, Jaffee EM et al (1995) A reassessment of the role of B7-1 expression in tumor rejection. J Exp Med 182:1415–1421

Editor-in-charge: Professor H. Gunicke

Structure and Function
of Eukaryotic Mono-ADP-Ribosyltransferases

Ian J. Okazaki and Joel Moss

Pulmonary-Critical Care Medicine Branch, National Heart, Lung,
and Blood Institute, National Institutes of Health, Bethesda, MD 20892, USA

Contents

1
Introduction

Mono-ADP-ribosylation is a post-translational modification of proteins in which the ADP-ribose moiety of NAD is transferred to an acceptor protein or amino acid. The reaction is catalyzed by NAD:(amino acid) protein ADP-ribosyltransferases of which several bacterial toxin ADP-ribosyltransferases have been well characterized (for reviews see Moss and Vaughan 1990). The bacterial toxin ADP-ribosyltransferases modify specific amino acids in effector proteins that regulate cellular metabolism. Cholera toxin and the related *Escherichia coli* heat-labile enterotoxin (LT) ADP-ribosylate an arginine residue in $G_s\alpha$, the stimulatory, α-subunit of the heterotrimeric GTP-binding (G) protein, resulting in the activation of adenylyl cyclase (Moss and Vaughan 1988). Pertussis toxin (PT) modifies a cysteine in the α-subunits of G_i, G_o and G_t resulting in uncoupling of the G-protein from its receptor (Ui 1990). Diphthamide, a post-translationally modified histidine in elongation factor-2 (EF-2), is ADP-ribosylated by diphtheria toxin (DT) and *Pseudomonas aeruginosa* exotoxin A (ETA), resulting in inhibition of protein synthesis and cell death (Wick and Iglewski 1990; Collier 1990). A group of clostridial ADP-ribosylating toxins (e.g., C2 toxin, iota toxin) modifies arginine 177 of actin, leading to inhibition of actin polymerization and breakdown of the microfilament network (Aktories 1994).

ADP-ribosyltransferase activity has been detected in numerous animal tissues including turkey erythrocytes (Moss et al. 1980), rat liver (Moss and Stanley 1981b), rabbit skeletal muscle (Soman et al. 1984), *Xenopus* tissues (Godeau et al. 1984), chicken heterophils (Tanigawa et al. 1984) and several murine cell lines (Soman et al. 1991); these transferases use arginine as an ADP-ribose acceptor. The turkey erythrocyte (Moss et al. 1980; Yost and Moss 1983), chicken heterophil (Tanigawa et al. 1984; Mishima et al. 1991), and rabbit skeletal muscle (Peterson et al. 1990; Zolkiewska et al. 1992) transferases have been purified and characterized, and transferase coding region cDNAs have been cloned from rabbit (Zolkiewska et al. 1992) and human (Okazaki et al. 1994) skeletal muscle, chicken heterophils (Tsuchiya et al. 1994), chicken nucleoblasts (Davis and Shall 1995) and mouse lymphocytes (Okazaki et al. 1996a, 1996b). Whereas poly(ADP-ribose) polymerase catalyzes the formation of branched ADP-ribose polymers attached to a number of nuclear proteins (Alvarez-Gonzales et al. 1994), the mono-ADP-ribosyltransferases modify a diverse group of substrates including $G_s\alpha$ (Duman et al. 1991), β/γ-actin (Mat-suyama and Tsuyama 1991), p33 (Mishima et al. 1991) and integrin $\alpha7$ (Zolkiewska and Moss 1993).

ADP-ribosylation of arginine appears to be a reversible process; free arginine can be regenerated from the ADP-ribosylated protein by the action of ADP-ribosylarginine hydrolases (for review, see Williamson and Moss 1990). Hydrolase activity was detected in the soluble fraction of turkey erythrocytes (Moss et al. 1985), cultured mouse cells (Smith et al. 1985) and rat skeletal muscle (Chang et al. 1986). ADP-ribosylarginine hydrolases have been purified from turkey erythrocytes and rat brain (Moss et al. 1988, 1992); the rat brain hydrolase cDNA was subsequently cloned (Moss et al. 1992). A regulatory eukaryotic ADP-ribosylation cycle was proposed based on the presence of ADP-ribosyltransferases and ADP-ribosylarginine hydrolases (Williamson and Moss 1990).

Regulation of nitrogen fixation by ADP-ribosyltransferases and ADP-ribosylarginine hydrolases has been shown to occur in the photosynthetic bacterium *Rhodospirillum rubrum* (reviewed by Ludden 1994). Dinitrogenase reductase, which is part of a nitrogen-reducing enzyme complex in *R. rubrum*, is inactivated by ADP-ribosylation of arginine 101 by dinitrogenase reductase ADP-ribosyltransferase (DRAT) in response to environmental stimuli such as darkness or a source of fixed nitrogen such as ammonia. Upon exposure of the bacteria to light, or depletion of the nitrogen source, the modified dinitrogenase reductase is activated by removal of the ADP-ribose group catalyzed by dinitrogenase reductase ADP-ribose glycohydrolase (DRAG), regenerating the free guanidino moiety of arginine. More recently, glutamine synthetase III from the symbiotic nitrogen fixing bacteria *Rhizobium meliloti*, was shown to be ADP-ribosylated at an arginine in vivo, resulting in inhibition of glutamine synthetase activity (Liu and Kahn 1995). The activity of the modified enzyme was restored by treatment with the turkey erythrocyte ADP-ribosylarginine hydrolase (Liu and Kahn 1995). Glutamine synthetase from the cyanobacterium *Synechocystis* sp. strain PCC 6803 was also inactivated by endogenous ADP-ribosylation (Silman et al. 1995). The activity of the modified enzyme could be restored by treatment with phosphodiesterase (Silman et al. 1995).

Endogenous ADP-ribosylation of cysteine was reported in human erythrocytes (Tanuma et al. 1987) and a 27-kDa, NAD:cysteine ADP-ribosyltransferase that modified G_i in erythrocyte and platelet membranes was purified (Tanuma et al. 1988). A human erythrocyte ADP-ribosylcysteine hydrolase that catalyzed the reverse reaction was also identified (Tanuma and Endo 1990), consistent with the presence of an ADP-ribosylation cycle involving cysteine residues in proteins. McDonald et al. (1992) subsequently reported the nonenzymatic formation of an ADP-ribosylthiazolidine from the reaction of ADP-ribose with L-cysteine, D-cysteine, cysteamine, and L-cysteine methyl ester, but not with dithiothreitol, β-mercaptoethanol, glutathione or N-acetyl-L-cysteine. The reaction was dependent upon the presence of free -NH$_2$ and -SH groups in the ADP-ribose acceptor. The thiazolidine linkage was NH$_2$OH- and HgCl$_2$-sensitive, whereas a thioglycoside ADP-ribosylcysteine(protein) formed by PT was sensitive to HgCl$_2$, but not to NH$_2$OH. The ADP-ribosylcysteine methyl-ester generated by incubation of the erythrocyte enzyme with cysteine methyl ester and NAD (Tanuma et al. 1988) may have been synthesized nonenzymatically. A two-step mechanism is possible, involving ADP-ribose generation by an NAD glycohydrolase followed by reaction of ADP-ribose with free cysteine, resulting in a thiazolidine product (McDonald and Moss 1994). Free ADP-ribose has been shown to react with a cysteine in actin to yield a product with the HgCl$_2$ and hydroxylamine sensitivity of ADP-ribosylated transducin formed by PT (Just et al. 1994). Similarly, free ADP-ribose reacted with aldehyde dehydrogenase to yield an ADP-ribosylated protein with NH$_2$OH and HgCl$_2$ sensitivity similar to that of the product of a PT-catalyzed reaction (McDonald and Moss 1993a). Further, proteins ADP-ribosylated on cysteine have been identified in rat tissue extracts (Jacobson et al. 1990). Chemical sensitivity alone, however, cannot distinguish enzymatic and nonenzymatic ADP-ribosylation of proteins (McDonald and Moss 1993a), and it remains to be determined whether endogenous ADP-ribose addition to cysteine in mammalian tissues is enzymatic or nonenzymatic.

It was previously reported that nitric oxide (NO) stimulated the ADP-ribosylation of glyceraldehyde 3-phosphate dehydrogenase (GAPDH) (Dimmeler et al. 1992, Kots et al. 1992). As the product was sensitive to HgCl$_2$, a cysteine appeared to be modified. Further analysis of the reaction product, however, established that the entire NAD molecule, not just ADP-ribose, was noncovalently bound to GAPDH (McDonald and Moss 1993b). Thus in this case as well, cysteine appears not to be enzymatically modified. It should be noted that in brain extracts, NO-dependent ADP-ribosyltransferases were observed, which modified both cysteine and arginine linked to solid supports (Schuman et al. 1994). Likewise, NO stimulated the ADP-ribosylation of cytosolic proteins from the mouse macrophage cell line,

ANA-1 (Sheffler et al. 1995), and NO-dependent ADP-ribosylation of β/γ-actin was demonstrated in human neutrophils (Clancy et al. 1995). The mechanism of these reactions is uncertain, pending purification of the products and the enzymes.

The purpose of this chapter is to characterize the mono-ADP-ribosyl-transferases and ADP-ribosylarginine hydrolases that have been purified and cloned from vertebrate species. In addition, recent evidence will be presented that demonstrates the existence of conserved regions in the amino acid sequences of viral, bacterial toxin and vertebrate ADP-ribo-syltransferases, which suggests a common mechanism of substrate binding and catalysis of the ADP-ribose transfer reaction among transferases.

2
Mono-ADP-Ribosyltransferases

2.1
Avian ADP-Ribosyltransferases

2.1.1
Turkey Erythrocyte ADP-Ribosyltransferases

Four distinct ADP-ribosyltransferases have been identified in turkey erythrocytes; transferases A (Moss and Vaughan 1978; Moss et al. 1980) and B (Yost and Moss 1983) were isolated from erythrocyte cytosol, transferase C from the plasma membrane, and transferase A' from the nucleus (West and Moss 1986). Transferase A, which was purified over 500 000-fold (Moss et al. 1980), catalyzes a reaction similar to that of cholera toxin (CT) (Moss and Vaughan 1978). The 28-kDa enzyme had a specific activity of 350 μmol-·min^{-1}·mg^{-1}using arginine methyl ester as the ADP-ribose acceptor, which was several orders of magnitude greater than that observed with CT. Like the stereospecific CT-catalyzed reaction, transferase A generated an α-anomeric ADP-ribosylarginine from β-NAD (Oppenheimer 1978; Moss et al. 1979). The kinetic mechanism of the turkey transferase, a rapid equilibrium random sequential model, was identical to that of CT; NAD and the ADP-ribose acceptor (agmatine) bind in random order to the transferase, with binding of one substrate having a negative effect on the subsequent binding of the other (Osborne et al. 1985).

Transferase A existed as a relatively inactive oligomer in the absence of NaCl (Moss et al. 1981). The oligomer dissociated, and enzyme activity increased more than tenfold, with the addition of up to 200 mM NaCl (Moss et al. 1981). Histones (20 µg/ml) also dissociated the transferase to an active monomer (Moss et al. 1982). Transferase activity was enhanced by chaotropic salts ($SCN^- > Br^- > Cl^- > F^- > PO_4^{3-}$) at low concentrations of ADP-ribose acceptor. Positively charged groups near the guanidino moiety of the substrate affected transferase activity, with arginine methyl ester > agmatine > arginine > guanidinopropionate > guanidine (Moss et al. 1981). The K_m for NAD at optimal concentrations of histone and NaCl was 15 µM (Moss et al. 1982); the K_m for arginine methyl ester in the presence of NaCl was 1.3 mM (Moss et al. 1981). NaCl and guanidino compounds also stimulated the NAD glycohydrolase activity of the transferase.

Histones and ovalbumin stabilized the monomeric transferase and served as ADP-ribose acceptors in vitro, although in the presence of agmatine, ADP-ribosylagmatine formation was favored over the formation of ADP-ribosylhistone (Moss and Stanley 1981a). Additionally, there was more than tenfold activation of the transferase at histone concentrations of 10–20 µg/ml, which were much lower than those required for histones to serve as ADP-ribose acceptors. In the absence of salt, ADP-ribosylagmatine formation in the presence of histone was 10 times that in the presence of ovalbumin. At high salt concentrations (300 mM NaCl), however, enzyme activity was not further stimulated by histone or ovalbumin (Moss et al. 1982).

Transferase A activity was enhanced approximately sixfold by lysophosphatidylcholine (Moss et al. 1984a); stimulation of transferase by lysophosphatidylcholine was significantly less than that by NaCl or histones. Unlike NaCl, however, lysophosphatidylcholine stabilized the enzyme against thermal denaturation. Both fatty acid ($C_{16} > C_{18} > C_{14} > C_{12} > C_{10}$ approximately C_8) and choline moieties were critical for activity; lysophosphatidylglycerol, lysophosphatidylserine, lysophosphatidylethanolamine, and lysophosphatidic acid had no effect on transferase activity. The apparent K_m for NAD was unchanged in the presence of NaCl and lysophosphatidylcholine, although the V_{max} in NaCl was approximately 1.7-fold that in lysophosphatidylcholine. The apparent K_m for agmatine was not different with NaCl, lysophosphatidylcholine, or NaCl plus lysophosphatidylcholine. These data are consistent with the hypothesis that lysophosphatidylcholine interacts directly with the transferase to stabilize the enzyme in an active conformation. Similarly, the nonionic detergents Triton X-100, Triton X-114, Tween 20 and Triton X-305 and the zwitterionic detergent, 3-[(cholamidopropyl)dimethylammonio]-1-propanesulfonate (CHAPS), enhanced transferase activity and protected against thermal inactivation. Maximal enzyme activation by CHAPS was less than that produced by NaCl, histones or lysophosphatidylcholine (Moss et al. 1984a).

ADP-ribosyltransferase activity was enhanced or inhibited by nucleoside triphosphates depending on the protein substrate (Watkins and Moss 1982). With lysozyme or soluble proteins from thymus as substrate, the rate of ADP-ribosylation was increased approximately 100% by 10 mM nucleoside triphosphates with ATP > ITP=GTP > CTP=UTP. ADP-ribosylation of histone $f_{2\alpha}$, however, was unaffected by ATP and inhibited by GTP.

The turkey erythrocyte transferase B, purified 270 000-fold, had an estimated molecular mass of 32 kDa (Yost and Moss 1983). This enzyme had apparent K_m values for NAD and arginine methyl ester of 36 μM and 3.0 mM, respectively. Transferase B, unlike transferase A, was inhibited approximately 40% by chaotropic salts and was not activated by histones. In addition, transferase B did not self-associate in the absence of NaCl or histone.

Transferases C and A' were partially purified from the particulate fraction of turkey erythrocytes (West and Moss 1986). Transferase C, a 26-kDa enzyme from the erythrocyte plasma membrane, had K_m values for NAD and agmatine of 15 μM and 2 mM, respectively, similar to those determined for transferases A and B. Transferase C, in contrast to those transferases, was unaffected by histones or salt (West and Moss 1986).

Transferase A', a 25-kDa protein localized to the erythrocyte nuclear fraction, differed from transferase A in its chromatographic behavior and subcellular localization. Like transferase A, transferase A' was stimulated by NaCl and histones. Transferase A', in contrast to the nuclear poly(ADP-ribose) polymerase, was insensitive to DNA and used simple guanidino compounds as ADP-ribose acceptors (West and Moss 1986).

2.1.2
In Vitro Substrates of Turkey Erythrocyte ADP-Ribosyltransferase A

The turkey transferase A catalyzed the ADP-ribosylation of a critical arginine in ovine brain glutamine synthetase which resulted in inhibition of synthetase activity (Moss et al. 1984b). Modification by transferase A of arginine 172 of *E. coli* glutamine synthetase similarly resulted in the parallel loss of glutamine biosynthetic activity and γ-glutamyltransferase activity (Moss et al. 1990). Activity of the modified glutamine synthetase was restored by treatment with the turkey erythrocyte ADP-ribosylarginine hydrolase (Moss et al. 1990).

The Ha-*ras* protooncogene product, p21 (Tsai et al. 1985), also served as a substrate for transferase A. Incubation of Ha-*ras* with enzyme and NAD in the presence of lysophosphatidylcholine, resulted in incorporation of approximately 3 mol of ADP-ribose per mol of p21 and decreased by approximately 50%, GTP-binding and GTPase activity (Tsai et al. 1985). Similarly, ADP-ribosylation of transducin by transferase A (Watkins et al. 1987) inhibited GTP-binding approximately 60%, and GTPase activity by approximately 90%. Although the α and β subunits of transducin were modified by transferase, inhibition of GTPase activity was predominantly due to ADP-ribosylation of the α-subunit.

Tubulin was ADP-ribosylated following incubation of cytosolic proteins from rat glioma cells with [^{32}P]NAD and CT (Hawkins and Browning 1982). Similarly, transferase A ADP-ribosylated the α and β polypeptide chains of chicken red blood cell tubulin with stoichiometries of 0.8–1.2 mol ADP-ribose per mol tubulin dimer (Raffaelli et al. 1992). Incubation of bovine brain extracts with turkey transferase A resulted in the incorporation of 2.4 mol of ADP-ribose per mol of tubulin and 30 mol of ADP-ribose per mol of high molecular weight microtubule-associated proteins (Scaife et al. 1992). ADP-ribosylation of tubulin inhibited microtubule assembly, and modification of assembled microtubules from bovine brain resulted in microtubule depolymerization.

Skeletal muscle α-actin, another in vitro substrate of transferase A, was ADP-ribosylated on arginine-95 and arginine-372 (Just et al. 1995a). Both monomeric G-actin and polymerized F-actin were ADP-ribosylated by transferase A. Modification of G-actin by the avian enzyme retarded monomer polymerization but did not alter the extent of F-actin formation. Actin-catalyzed ATPase activity was likewise unaffected by transferase A-catalyzed ADP-ribosylation. *Clostridium perfringens* iota toxin preferentially ADP-ribosylated G-actin, resulting in the formation of an F-actin capping protein (Wegner and Aktories 1988). Modification, which inhibited polymer formation (Aktories et al. 1986) and actin ATPase activity, was on arginine 177 (Vandekerckhove et al. 1987; Aktories 1994).

2.1.3
Chicken Heterophil ADP-Ribosyltransferases

An ADP-ribosyltransferase, purified 219-fold from chicken heterophil granules, had a specific activity of 0.4 mmol·mg^{-1}·h^{-1}(Mishima et al. 1991); it
was subsequently found to be identical to the transferase previously isolated from hen liver (Tanigawa et al. 1984; Mishima et al. 1991). The molecular mass of the purified transferase was estimated at 27.5 kDa (Mishima
et al. 1991); a second 28.0 kDa isoform was recently separated on sodium
dodecyl sulfate-polyacrylamide gel electrophoresis (SDS-PAGE) (Yamada
et al. 1994). The 27.5 kDa transferase utilized histones, casein, protamine,
and simple guanidino compounds (e.g., arginine methyl ester, agmatine)
as ADP-ribose acceptors (Tanigawa et al. 1984). The Km values for NAD
with arginine methyl ester, histone H1, and histone H2a as acceptors were
0.07, 0.56 and 0.29 mM, respectively, all of which were higher than that
for the turkey erythrocyte transferase; values for arginine methyl ester
and agmatine were 24 and 1.9 mM, respectively.

The chicken transferase exists as an active monomer. Unlike turkey
transferase A, the heterophil transferase was inhibited by NaCl and lysophosphatidylcholine; CHAPS and Triton X-100 had no effect on enzyme
activity. The heterophil enzyme was stimulated by sulfhydryl reagents such
as β-mercaptoethanol and polyanions such as double-stranded DNA, RNA
or poly(L-glutamate) (Mishima, et al. 1989). Like the turkey transferase,
the heterophil enzyme had NAD glycohydrolase activity. Nicotinamide
release from NAD exceeded ADP-ribosylation with all ADP-ribose acceptors tested; the ratio of nicotinamide release to ADP-ribosyltransferase
activity varied with the ADP-ribose acceptor (Tanigawa et al. 1984). These
data are consistent with the fact that all ADP-ribosyltransferases are, in
addition, NAD glycohydrolases. In contrast to the turkey enzyme, however,
the heterophil transferase was significantly auto-ADP-ribosylated when
incubated with [^{32}P]NAD in a zymographic analysis (Yamada et al. 1994).

Using degenerate oligonucleotide primers based on amino acid sequences of proteolytic fragments of the purified heterophil transferase,
two similar but distinct transferases, AT1 and AT2, were cloned from a
chicken bone marrow cDNA library (Tsuchiya et al., 1994). The AT1 cDNA
hybridized with a 1.5-kb band on a Northern blot of total RNA from
chicken bone marrow but not peripheral heterophils, spleen, liver, brain,
lung, heart or skeletal muscle. Oligonucleotide probes specific for AT1 and
AT2 hybridized on Northern blot to mRNAs of similar size. AT1 and AT2

had open reading frames of 312 amino acids with 78.3% identity. In the coding region, there was 89.4% nucleotide sequence identity between AT1 and AT2, whereas identities in the 5'- and 3'-untranslated regions were 100% and 98.4%, respectively. The hydrophobicity plot of the deduced amino acid sequence of AT1, along with direct N-terminal sequencing of purified transferase, was consistent with the presence of a hydrophobic amino terminal signal peptide. Further, amino acid sequencing at the C-terminus of AT1, ended at glutamine 266, consistent with processing of the amino and carboxy termini to yield the mature enzyme.

The coding region cDNAs were used to express AT1 and AT2 activity in COS 7 cells. AT1 activity was detected in the culture medium, whereas AT2 activity was found in both medium and cell lysate. AT1 activity was thiol-dependent and was optimal with 5 mM β-mercaptoethanol, whereas 200 mM NaCl inhibited activity approximately 70%. In contrast, AT2 activity did not require thiol, although activity was enhanced 100% by 5 mM β-mercaptoethanol. NaCl stimulated AT2 activity 80% above that with β-mercaptoethanol alone. Whether AT2 corresponds to the 28-kDa isoform purified from the heterophil granules has not yet been resolved (Tsuchiya et al. 1994; Yamada et al. 1994).

2.1.4
Other Chicken ADP-Ribosyltransferases

Endogenous ADP-ribosyltransferase activity was demonstrated in chicken spleen membranes (Obara et al. 1991). GTP-dependent ADP-ribosylation of Gsα resulted in increased adenylyl cyclase activity (Obara et al. 1991), similar to that produced by CT.

More recently, an enzyme with ADP-ribosyltransferase activity was released from chicken spleen membranes treated with phosphatidylinositol-specific phospholipase C (PI-PLC), but not by 1 M NaCl or 1% Triton X-100, consistent with the transferase being a glycosylphosphatidylinositol (GPI)-linked enzyme (Tsuchiya et al. 1995). The molecular sizes of the transferase were estimated to be 42 and 46 kDa by an in situ, zymographic ADP-ribosyltransferase assay under nonreducing conditions. In contrast to the chicken heterophil transferases AT1 and AT2, the transferase from chicken spleen membranes was not affected by 5 mM DTT or 200 mM NaCl (Tsuchiya et al. 1995).

An ADP-ribosyltransferase was recently cloned from a chicken erythroblast cDNA library using the radiolabeled rabbit skeletal muscle trans-

ferase cDNA as a probe (Davis and Shall 1995). The open reading frame
of the cloned gene encodes a 300-amino acid protein that is 50%–52%
identical to the skeletal muscle and two heterophil transferases. The
chicken erythroblast transferase cDNA has not been expressed nor has
it been determined whether the erythroblast transferase is homologous
to the turkey erythrocyte enzyme or that detected in chicken spleen membranes.

2.1.5
In Vitro Substrates of the Heterophil Transferase

Histones ADP-ribosylated by the heterophil transferase served as initiators
for poly(ADP-ribose) synthesis catalyzed in vitro by purified poly(ADP-
ribose) polymerase (Tanigawa et al. 1984). Whether ADP-ribosylation of
histone proteins by a heterophil granule transferase occurs in vivo is unknown.

The in vitro ADP-ribosylation of several proteins by the purified heterophil transferase inhibited subsequent phosphorylation of the modified
substrate. ADP-ribosylation of arginine 34 of histone H1 by the transferase
suppressed phosphorylation of serine 38 by cAMP-dependent protein kinase (Tanigawa et al. 1983a, b; Ushiroyama et al. 1985). Likewise, ADP-ribosylation of arginine residues in the α and β subunits of phosphorylase
kinase blocked cAMP-dependent phosphorylation of the enzyme, thus inhibiting activation of the kinase (Tsuchiya et al. 1985). Inactivation of L-type
pyruvate kinase by cAMP-dependent phosphorylation was blocked by
transferase-catalyzed ADP-ribosylation (Matsuura et al. 1988).

The bifunctional enzyme 6-phosphofructose-2-kinase/fructose-2,6-bis-
phosphatase (PFK-2/FBPase-2), which is involved in the synthesis and degradation of fructose 2,6-bisphosphate, was ADP-ribosylated by the heterophil transferase on arginine at positions 29 and 30 (Rosa et al. 1995).
ADP-ribosylation of PFK-2/FBPase-2 decreased PFK-2 activity without altering FBPase-2 activity and blocked phosphorylation of serine 32, catalyzed by cAMP-dependent protein kinase (PKA). On the other hand, PKA-
catalyzed phosphorylation of PFK-2/FBPase-2 inhibited PFK-2 activity but
activated FBPase-2. Phosphorylation of PFK-2/FBPase-2, however, did not
suppress subsequent ADP-ribosylation by the chicken transferase. These
data suggest that PFK-2 activity could be regulated by a balance between
phosphorylation and ADP-ribosylation (Rosa et al. 1995).

Another major in vitro ADP-ribose acceptor for the heterophil transferase was a 33-kDa protein, p33, which was purified from heterophil granules (Mishima et al. 1991). In permeabilized heterophils incubated with [^{32}P]NAD, p33 was the predominant ADP-ribosylated protein. Modification of p33 by the heterophil transferase resulted in the incorporation of 4 mol of ADP-ribose per mol of p33 (Mishima et al. 1991). Amino acid sequences of purified peptide fragments of p33 exhibited similarities to that of the *myb*-induced myeloid protein-1, *mim-1* (Yamada et al. 1992), which appears to be developmentally regulated and is expressed in heterophil precursors (Ness et al. 1989).

Nonmuscle β/γ-actin, skeletal muscle α-actin, and smooth muscle γ-actin were ADP-ribosylated in vitro by the heterophil transferase (Terashima et al. 1992). Further, incubation of permeabilized heterophils with [^{32}P]NAD, resulted in labeling of β/γ-actin. ADP-ribosylation completely inhibited polymerization of the actin isoforms, similar to the effect seen with Clostridium perfringens iota toxin (Schering et al. 1988). Unlike the iota toxin, however, the heterophil enzyme also modified F-actin. It was subsequently determined that the heterophil transferase modified arginine 28 and arginine 206 in G-actin but only arginine 28 in F-actin (Terashima et al. 1995). ADP-ribosylation of arginine 28 and arginine 206, unlike modification of arginine 177 by iota toxin, did not affect intrinsic ATPase activity of actin. Modification of arginine 206, however, inhibited interaction of G-actin with deoxyribonuclease (DNase) I and affected actin polymerization. Arginine 206 is located on the pointed end of the actin molecule and is near a minor contact point between actin and DNase I. Arginine 28, on the other hand, is located on the outer surface of actin and does not interfere with the interaction of G-actin with DNase I, or G-actin binding to myosin subfragment-1 (Terashima et al. 1995). Endogenous ADP-ribosylation of heterophil nonmuscle actin was proposed as a mechanism for modulating heterophil functions dependent on actin polymerization and depolymerization (Terashima et al. 1992).

2.2.
Mammalian ADP-Ribosyltransferases

2.2.1
Skeletal and Cardiac Muscle ADP-Ribosyltransferases

An ADP-ribosyltransferase that utilized guanidino compounds as ADP-ribose acceptors was partially purified from rabbit skeletal muscle (Peterson et al. 1990). In subsequent studies, the 36-kDa muscle transferase was extensively purified (215 000-fold) and exhibited a specific activity, at optimal NAD concentration (2 mM), of 68 µmol'min^{-1}'mg^{-1} with agmatine as ADP-ribose acceptor (Zolkiewska et al. 1992), a value similar to that of the turkey transferases (Moss et al. 1980; Yost and Moss 1983).

Oligonucleotide primers, based on amino acid sequences of tryptic peptides, were utilized to clone the rabbit transferase from a skeletal muscle cDNA library (Zolkiewska et al. 1992). A rabbit muscle transferase-specific oligonucleotide probe hybridized, on a Northern blot of total RNA from rabbit tissues, with a 3-kb mRNA from skeletal and cardiac muscle, but did not hybridize with mRNA from smooth muscle, brain, lung, kidney, spleen or liver (Zolkiewska et al. 1992), demonstrating apparent tissue-specific expression of the muscle transferase.

The human skeletal muscle ADP-ribosyltransferase was cloned by PCR-based procedures from human skeletal muscle poly(A)$^+$ RNA (Okazaki et al. 1994). Cross-species conservation of transferase expression was demonstrated on Northern blot of total RNA from skeletal and cardiac muscle, where a PCR-generated human muscle transferase cDNA probe hybridized with a 1.2-kb band from mouse and rat, a major 3.0-kb and minor 4.0-kb band from rabbit, a major 3.8-kb band and minor 5.7-kb band from monkey and a 5.7-kb band from human skeletal muscle. Further, polyclonal anti-rabbit skeletal muscle transferase antibodies reacted on immunoblots with a 36-kDa protein from bovine, dog and rabbit cardiac muscle and a 40-kDa protein from human skeletal muscle (Okazaki et al. 1994) supporting the conclusion that the structure of the muscle transferases is conserved across species.

The hydrophobicity profile of the deduced amino acid sequence of the rabbit skeletal muscle transferase revealed hydrophobic amino and carboxy termini consistent with the presence of a GPI anchor (Zolkiewska et al. 1992). The sequences of the signal peptides were similar to those used for transport and processing of nascent proteins prior to attachment

of a GPI-anchor (Gerber et al. 1992). Expression of the rabbit muscle transferase cDNA in rat mammary adenocarcinoma (NMU) cells demonstrated
that the majority of transferase activity was localized to the membrane
fraction and could be solubilized by incubating intact cells with PI-PLC,
a characteristic consistent with a GPI-linked enzyme (Okazaki et al. 1994).
In addition, a 36-kDa protein released by PI-PLC from transformed, but
not control, NMU cells reacted with anti-transferase antibodies as well as
with anti-cross-reacting determinant (CRD) antibodies, which recognize
the oligosaccharide inositol-1,2-cyclic phosphate moiety exposed on GPI-
anchored proteins after cleavage with PI-PLC. NMU cells transformed with
a cDNA lacking the carboxy-terminal signal peptide, secreted transferase
activity into the medium, presumably due to absence of the GPI-anchoring
sequence. Partially purified transferases from rabbit and human skeletal
muscle reacted with anti-CRD antibodies after incubation with PI-PLC,
consistent with the hypothesis that the ADP-ribosyltransferase was GPI-
anchored in native tissues (McMahon et al. 1993; Okazaki et al. 1994).
Although a skeletal muscle transferase exists as an exoenzyme (Zolkiewska
and Moss 1993), transferase activity was also found associated with the
sarcoplasmic reticulum (Soman et al. 1984) and cytoplasmic face of the
sarcolemma and transverse tubule membranes, isolated by fractionation
on sucrose gradients (Klebl et al. 1994).

ADP-ribosyltransferase activity was detected in C2C12 and G8 mouse
skeletal muscle cell lines (Zolkiewska and Moss 1993). Enzyme activity
increased upon differentiation of myoblasts to myotubes and was released
from intact cells by treatment with PI-PLC. Incubating fractions from
muscle cell lysates with radiolabeled NAD resulted in the ADP-ribosylation
of several proteins (Piron and McMahon 1990; Klebl et al. 1994), whereas
when intact C2C12 cells were used, a 97-kDa membrane protein was the
major ADP-ribosylated product (Zolkiewska and Moss 1993). The radiolabel associated with the 97-kDa protein was released by NaOH and
NH_2OH, consistent with an ADP-ribose-arginine linkage. ADP-ribosylation
of the 97-kDa protein was inhibited by treatment of cells with PI-PLC
before, but not after, incubation with $[^{32}P]NAD$. Mobilities of the radiolabeled protein on SDS-PAGE under reducing and nonreducing conditions
(consistent with 97-kDa and 140-kDa proteins, respectively), were similar
to those reported for integrin $\alpha 7$. Using a laminin affinity column, the
labeled protein was purified 150-fold. Amino acid sequences of the N-terminal and internal peptide fragments were identical to that of an isoform
of rat integrin $\alpha 7$ (Zolkiewska and Moss 1993).

By limited trypsin digestion of labeled integrin α7 modified at 5 µM NAD, the site of ADP-ribosylation was narrowed to a 39-kDa segment located between the ligand-binding and transmembrane domains that includes amino acids 575 to 886 (Zolkiewska and Moss 1993). At higher NAD concentrations (75 µM), integrin α7 was modified also in the 63-kDa N-terminal segment; ADP-ribosylation did not detectably affect α7β1 heterodimer formation or its association with the cytoskeleton or laminin (Zolkiewska and Moss 1995). ADP-ribosylated integrin α7 was rapidly hydrolyzed by extracellular phosphodiesterases, yielding phosphoribosyl-integrin and 5'AMP. Cells containing processed integrin (phosphoribosyl-or ribosyl-integrin, generated from phosphoribosyl-integrin by an extracellular phosphatase), were refractory to further ADP-ribosylation for at least 1 h. The studies are consistent with the conclusion that the cytosolic ADP-ribosylarginine hydrolase may not be involved in processing the extracellular ADP-ribosylated integrin (Zolkiewska and Moss 1995). Expression of the ADP-ribosyltransferase in parallel with that of integrin α7 during myogenesis (Song et al. 1992), and the demonstration of ADP-ribosylation of integrin α7 in intact cells suggested a regulatory role for such a modification in muscle cell development. Furthermore, inhibition of proliferation and differentiation of embryonic chick myoblasts by *meta*-iodobenzylguanidine in vitro, an inhibitor of arginine-specific mono-ADP-ribosyltransferases (Khardia et al. 1992), was consistent with the hypothesis that ADP-ribosylation may, to some extent, modulate myogenesis.

2.2.2
Brain ADP-Ribosyltransferases

Four distinct 66-kDa ADP-ribosyltransferases were purified approximately 3000-fold from rat brain (Matsuyama and Tsuyama 1991). Using β/γ-actin as ADP-ribose acceptor, the K_m values for NAD were 17.2 µM, 12.1 µM, 24.6 µM, and 30.6 µM for transferases I, II, III, and IV, respectively.

Phospholipids had different effects on the activity of the brain transferases (Matsuyama and Tsuyama 1991). Transferase I was activated by lysophosphatidylcholine, phosphatidylcholine, phosphatidylethanolamine and phosphatidylserine, whereas transferase II was stimulated by lysophosphatidylcholine, and transferase III by lysophosphatidylcholine, phosphatidylethanolamine and phosphatidylserine; transferase IV was inhibited by phospholipids. Activity of the brain transferases was also affected by soluble ADP-ribosylation factors (sARF) I and II, 20-kDa guaninine nucleotide-binding proteins from bovine brain, which enhance in a GTP-dependent manner, cholera-toxin catalyzed ADP-ribosylation and are involved in regulation of vesicular transport (Welsh et al. 1994; Moss and Vaughan 1995). sARF I activated transferase II, whereas sARF II enhanced the activity of transferases I and IV. Transferase III activity was suppressed, however, by both sARFs (Matsuyama and Tsuyama 1991). ADP-ribose and ADP, but not nicotinamide, inhibited the brain transferases.

Nonmuscle β/γ-actin and smooth muscle γ-actin was ADP-ribosylated by all four transferases, whereas skeletal muscle α-actin was a substrate for only transferases I and IV. Modification of nonmuscle G-actin by transferase inhibited actin polymerization in vitro. Microtubule-associated protein (MAP-2), critical for microtubule assembly and disassembly, was also ADP-ribosylated by the four transferases (Takenaka et al. 1994). Furthermore, the brain transferases modified in vitro G_s and G_o, the heterotrimeric GTP-binding proteins that are ADP-ribosylated by PT (Sternweis and Robishaw 1984; Matsuyama and Tsuyama 1991). Transferases I, II, and IV modified G_s, and transferases I, III and IV modified G_o. None of the transferases, however, ADP-ribosylated histones (Matsuyama and Tsuyama 1991).

Incubation of rat brain homogenates with [^{32}P]NAD and 5'-guanylyl-imidodiphosphate [Gpp(NH)p] resulted in the ADP-ribosylation of several proteins, including 20-, 42-, 45- and 50-kDa proteins (Duman et al. 1991). Under similar conditions, CT ADP-ribosylated the 20-, 42- and 45-kDa proteins, although toxin-catalyzed modification was considerably greater than that which occurred endogenously. Endogenous and toxin-catalyzed ADP-ribosylation was enhanced by isoniazid and 3-acetylpyridine adenine dinucleotide, which are inhibitors of NAD-glycohydrolase activity (Tamir and Gill 1988), and by Triton X-100. ADP-ribosyltransferase activity was detected in the particulate, but not in the soluble, fraction of the cerebral cortex. The solubility of transferase activity was not affected by the addition of Triton X-100. The extent of ADP-ribosylation in the particulate fraction was much less than that in the crude homogenate. Addition of the soluble to the particulate fraction, however, restored transferase activity to the level found in the crude homogenate. Increased ADP-ribosylation of the 20-, 42-, 45- and 50-kDa proteins was observed in brain homogenates from rats that had been treated with corticosterone for seven days (Duman et al. 1991).

There were significant regional differences in ADP-ribosylation in the rat brain. The highest levels of ADP-ribosylation of the 42- and 45-kDa proteins were found in the hippocampus, hypothalamus and cerebral cortex, with intermediate levels in the midbrain, thalamus and neostriatum; the lowest levels were observed in the cerebellum. Modification of the 50-kDa protein was similar throughout the brain except for significantly higher levels of ADP-ribosylation in the neostriatum (Duman et al. 1991).

One dimensional peptide maps of the labeled 42- and 45-kDa proteins were identical to those of the two major forms of $G_s\alpha$. The 20-kDa protein was felt to be ADP-ribosylation factor (ARF) based on its modification by CT, and increased endogenous ADP-ribosylation after chronic administration of glucocorticoids (Duman et al. 1991). Glucocorticoids have been shown to increase the expression of $G_s\alpha$ (Saito et al. 1989) and ARF (Duman et al. 1990) in the cerebral cortex. The identity of the 50-kDa protein has yet to be determined.

Addition of LiCl (0.3–1 mM), at therapeutic concentrations, to extracts of rat frontal cortex inhibited ADP-ribosylation of proteins, including $G_s\alpha$ and a 48-kDa protein, thought to be the 43-kDa growth-associated protein (GAP-43) (Nestler et al. 1995). Inhibition of ADP-ribosylation was not observed with NaCl or KCl. Li^+ also inhibited ADP-ribosylation in brain extracts, in the presence of sodium nitroprusside, a NO donor which stimulates endogenous ADP-ribosylation (Duman et al. 1991). In the frontal cortex of rats treated with LiCl for 4 weeks, however, ADP-ribosylation was increased. In contrast, short-term (6 days) treatment with LiCl had no effect on endogenous ADP-ribosylation despite therapeutic serum Li^+ levels (Nestler et al. 1995). The clinical consequences of alteration of ADP-ribosylation in the brain by Li^+ administration remain to be determined.

Other proteins isolated from brain can also be substrates of endogenous ADP-ribosyltransferases. The 80–90-kDa myristoylated, alanine-rich, protein kinase C substrate protein (MARKS) (Chao et al. 1995), and the growth-associated protein, B-50/GAP-43 (Coggins et al. 1993a) and a B-50/GAP-43-related protein, neurogranin (Coggins et al. 1993b), was modified in brain homogenates although it is not known whether ADP-ribosylation alters the function of these proteins.

The CA-1 region of the rat hippocampus exhibits a NO-stimulated ADP-ribosyltransferase activity (Schuman et al. 1994). In the presence of hippocampal homogenates, arginine and cysteine linked to cyanogen bromide-activated Sepharose beads served as ADP-ribose acceptors. Moreover, inhibitors of mono-ADP-ribosyltransferases, phylloquinone (100 μM) and nicotinamide (1–10 mM), were able to block long-term potentiation in hippocampal slices without affecting basal excitatory synaptic transmission, inhibitory postsynaptic potential, or N-methyl-D-aspartate (NMDA) receptor-mediated transmission. ADP-ribosylation of the α-subunit of G-proteins or B-50/GAP-43 may play a role in neurotransmitter release affecting long-term potentiation (Schuman et al. 1994).

ADP-ribosylation of several proteins in brain and retina was altered in alloxan diabetic rats. A marked decrease in PT-catalyzed ADP-ribosylation of retinal $G_{i\alpha}$ and/or $G_{o\alpha}$ was demonstrated 5 weeks after alloxan injection (Abbracchio et al. 1991), consistent with an increase in endogenous ADP-ribosylation of retinal proteins. There was a concomitant increase in cAMP production felt to be secondary to unopposed G_s activity. An increase in endogenous modification of G_i/G_o in the brain of diabetic rats was also evident 14 weeks after alloxan injection. This decrease in PT-catalyzed [^{32}P]ADP-ribosylation was not due to reduced levels of α-subunits of G proteins or their mRNA transcripts, the substrates of the transferase enzyme (Finco et al. 1992). Additionally, the increased endogenous ADP-ribosylation of several cytosolic and membrane proteins in the retina of diabetic rats was normalized by inhibiting endogenous mono-ADP-ribosyltransferases by treatment of the rats with sylbin, a flavanoid with ADP-ribosyltransferase inhibitory activity (Donadoni et al. 1995). Sylbin also prevented the reduced sciatic nerve axonal transport of substance P that has been observed in diabetic rats. Sylbin had no affect on hyperglycemia or weight loss due to diabetes (Donadoni et al. 1995). Insulin selectively normalized the endogenous ADP-ribosylation of a 38-kDa cytosolic and a 39-kDa membrane protein in retina, in addition to improving hyperglycemia and weight loss of the diabetic rats (Gorio et al. 1995). Several other retinal proteins from diabetic rats that demonstrated increased endogenous ADP-ribosylation were unaffected by insulin.

An ADP-ribosyltransferase from the retinal rod outer segment was regulated by the NO donor SIN-1 (Ehret-Hilberer et al. 1992). NOsynthase purified from bovine retinal rod inner segments enhanced ADP-ribosylation of rod outer segment proteins, including $G_{t\alpha}$, consistent with the regulation of photoreceptor signal transduction by NO (Zoche and Koch 1995). The presence of brain ADP-ribosyltransferases that modify a variety of endogenous proteins, including actin and GTP-binding proteins, is consistent with the notion that there is a critical role for ADP-ribosylation in neuronal function.

2.2.3
Lymphocyte ADP-Ribosyltransferases

ADP-ribosyltransferase activity has been reported in murine lymphoid cells including T cell lymphomas (Yac-1), T cell hybridomas, and a thymoma cell line (Soman et al. 1991). A correlation between the ADP-ribosylation of membrane proteins and cytotoxic T cell function has been described (Wang et al. 1994). Incubation of murine cytotoxic T lymphocytes (CTL) with 10 μM NAD inhibited CTL proliferation in a mixed lymphocyte reaction, whereas 100 μM NAD partially suppressed cytolytic activity. Inhibition of CTL proliferation and cytotoxicity by NAD was associated with inhibition of CTL-target cell binding. NAD did not affect T cell receptor-mediated signaling. Incubation of CTL with [^{32}P]NAD resulted in the labeling of numerous membrane proteins. [^{32}P]ADP-ribosylation was not inhibited by 100-fold molar excess of the NAD metabolites, nicotinamide, ADP-ribose or cyclic-ADP-ribose, 5'AMP, adenosine or nicotinamide mononucleotide. These metabolites likewise had no effect on CTL proliferation or cytotoxicity. The radiolabel was released from proteins by NaOH or NH$_2$OH, but not HgCl$_2$, consistent with an ADP-ribose-arginine linkage. An approximately 35-kDa protein with guanidine-specific ADP-ribosyltransferase activity was released during incubation of intact CTL with PI-PLC. In addition, PI-PLC-treatment of cells resulted in the partial loss of the inhibitory effect of NAD on CTL proliferation, and totally eliminated the suppressive effect of NAD on cytotoxicity (Wang et al. 1994). ADP-ribosylation of a 40-kDa CTL membrane protein (p40) resulted in the inhibition of p56^{lck}, a tyrosine kinase that exists in a complex with p40 (Wang et al., 1996). Further, release of the membrane-bound transferase following treatment with PI-PLC prevented the NAD-induced suppression of kinase activity. The relationship between the inhibition of p56^{lck} following ADP-ribosylation of p40 and the inhibition of CTL proliferation has not been established.

The molecular structure of a lymphocyte transferase (Yac1) cloned from Yac-1 cells appears to be quite similar to its muscle counterparts (Okazaki et al. 1996a). A second lymphocyte transferase from Yac-1 cells (Yac2) has been cloned and characterized (Okazaki et al. 1996b). The deduced amino acid sequences of the Yac1 and Yac2 transferases were only 32% identical and the sequence of the Yac2 transferase was only approximately 30% identical to those of the rabbit skeletal muscle and chicken heterophil transferases, and the rat T cell alloantigen RT6.2, an NAD glycohydrolase.

Unlike the Yac1 transferase, Yac2 had greater NAD glycohydrolase than NAD:arginine ADP-ribosyltransferase activity. With agmatine as the ADP-ribose acceptor, the K_m values for NAD of the Yac1 and Yac2 transferases were 118 μM and 142 μM, respectively; those for agmatine were 9.4 mM and 15 mM. NMU cells transformed with the Yac2 cDNA demonstrated cell surface transferase and NADase activities that were not released by PI-PLC. On Northern analysis, a Yac1 cDNA probe hybridized with a major 1.6-kb band and a minor 7.0-kb band in poly(A)$^+$ RNA from mouse cardiac and skeletal muscle, consistent with the approximately 70% identity of the nucleotide and deduced amino acid sequences of the Yac1 and rabbit skeletal muscle transferases (Okazaki et al. 1996a). A Yac2 cDNA probe hybridized with 1.6- and 2.0-kb bands in poly(A)$^+$ RNA from mouse testis. There was also weak hybridization with a 1.6-kb mRNA from mouse heart and skeletal muscle (Okazaki et al. 1996b). It is not yet known whether the Yac1 and/or the Yac2 transferases play a role in modulating CTL function.

CD38 is a differentiation antigen on the surface of lymphocytes that catalyzes the hydrolysis of NAD to ADP-ribose and also generates small quantities of cyclic ADP-ribose (Howard et al. 1993). Cyclic ADP-ribose, a potent second messenger, has the ability to mobilize intracellular stores of calcium by an inositol 1,4,5-trisphosphate-independent mechanism (Lee et al. 1989). The extracellular domain of CD38 was expressed in the baculovirus vector, generating a soluble CD38 (sCD38) protein. Lysozyme, interleukin-2 and myoglobin were ADP-ribosylated on cysteines in the presence of sCD38 and NAD; sCD38 was also auto-ADP-ribosylated (Grimaldi et al. 1995). ADP-ribose was released from the modified lysozyme by $HgCl_2$, but not by NH_2OH, consistent with the presence of a thioglycosidic bond. Lysozyme was modified by sCD38-generated ADP-ribose at a rate similar to that occuring in the presence of NAD and sCD38, consistent with a nonenzymatic ADP-ribosylation of cysteine (Grimaldi et al. 1995).

2.2.4
Other Mammalian ADP-Ribosyltransferases

Several proteins including $G_s\alpha$ were modified by an endogenous ADP-ribosyltransferase in permeabilized NG108-15 cells, a neuroblastoma-glioma hybrid (Donnelly et al. 1992). Incubation of permeabilized cells with 50 mM nicotinamide, 25 mM benzamide, or 10 mM 5-bromo-2'-deoxyuridine, inhibitors of ADP-ribosyltransferases, reduced ADP-ribosyl-ation of all proteins. In cells incubated with nicotinamide for 18 h, however, there was an increase in membrane-associated $G_s\alpha$ resulting in a significant increase in adenylyl cyclase activity. Endogenous ADP-ribosylation of $G_s\alpha$ and subsequent stimulation of adenylyl cyclase was similarly demonstrated in cardiac muscle membranes (Feldman et al. 1987; Quist et al. 1994), platelets (Molina y Vedia et al. 1989) and chicken spleen membranes (Obara et al. 1991).

Thirty-eight- and 50-kDa cytosolic proteins from rat basophilic leukemia (RBL) and Fischer rat thyroid line 5 (FRTL5) cells were ADP-ribosylated during incubation of cellular extracts with [^{32}P]NAD (De Matteis et al. 1994). ADP-ribosylation of these proteins was stimulated by the fungal toxin brefeldin A (BFA), which inhibits constitutive protein secretion, and disrupts the structure and function of organelles involved in protein trafficking (Pelham 1991; Klausner et al. 1992). The concentrations of BFA that stimulated ADP-ribosylation were similar to those that inhibited ADP-ribosylation factor binding to Golgi membranes (Tsai et al. 1993; De Matteis et al. 1994). The ADP-ribose-protein linkage was sensitive to treatment with HCl and NaOH but stable to NH_2OH and $HgCl_2$ which is unlike that formed by the known arginine-, cysteine-, and asparagine-specific transferases (De Matteis et al. 1994), and also from the linkage that results from nonenzymatic ADP-ribosylation of cysteine (McDonald and Moss 1993b). The 38-kDa protein was identified as glyceraldehyde-3-phosphate dehydrogenase by its electrophoretic behavior on two-dimensional gels. The 50-kDa protein BARS-50, a substrate for BFA-stimulated ADP-ribosylation, is a novel GTP-binding protein (Di Girolamo et al. 1995). Although GTP and $G_{\beta\gamma}$ were able to inhibit subsequent ADP-ribosylation of BARS-50, BARS-50 did not react with anti-G_α-specific antibodies (Di Girolamo et al. 1995).

NO stimulates the ADP-ribosylation of several proteins in a variety of cellular systems. NO released from 1,1-diethyl-2-hydroxy-2-nitrosohydrazine (DEA/NO) stimulated the ADP-ribosylation of three cytosolic proteins (28, 33 and 39 kDa) from the mouse macrophage cell line ANA-1 (Sheffler et al. 1995). NO generated after interferon-γ and lipopolysaccharide induction of NO synthase in ANA-1 cells similarly enhanced the ADP-ribosylation of the 39-kDa protein. Incubation of membrane and cytosolic proteins with CT resulted in the ADP-ribosylation of numerous proteins including 28-kDa and 33-kDa proteins, but not the 39-kDa substrate. PT, on the other hand, modified only 41- and 29-kDa proteins (Sheffler et al.

1995). The identities of the NO-induced ADP-ribosylated substrates and the effects of the modification on macrophage function have not been determined.

NO-dependent ADP-ribosylation of β/γ-actin was demonstrated in human neutrophils (Clancy et al. 1995). The factor responsible for ADP-ribosylation was localized to the plasma membrane fraction. Using nicotinamide- or adenine-labeled NAD, it was demonstrated that the NO-mediated ADP-ribosylation of actin was due to incorporation of the adenine, but not the nicotinamide, moiety. The NO-induced modification of G-actin resulted in transient inhibition of actin polymerization and neutrophil adherence to fibronectin-coated surfaces (Clancy et al. 1995). The chemical sensitivity of the ADP-ribose-actin bond, however, was not determined.

The NO-induced ADP-ribosylation of actin in neutrophil lysates (Clancy et al. 1995) is apparently distinct from the nonenzymatic ADP-ribosylation described in neutrophils and platelets (Just et al. 1994). Incubation of cell lysates from human platelets and polymorphonuclear leukocytes (PMN), with [^{32}P]NAD, resulted in the [^{32}P]ADP-ribosylation of β/γ-actin, but not skeletal muscle α-actin. The cell lysates were modified to a greater extent in the presence of [^{32}P]ADP-ribose than [^{32}P]NAD. Dithiothreitol increased incorporation of ADP-ribose, whereas NO released from SIN-1 had no effect. The endogenously formed ADP-ribose-actin bond was sensitive to HgCl$_2$, but not NH$_2$OH, consistent with a thioglycosidic bond. Phalloidin-induced polymerization of the modified actin was inhibited in a manner similar to that observed after ADP-ribosylation of actin by *Clostridium perfringens* iota toxin or *Clostridium botulinum* C2 toxin. Conversely, prior polymerization of actin blocked the subsequent nonenzymatic ADP-ribosylation of actin (Just et al. 1994).

2.3
ADP-Ribosyltransferase and NAD Glycohydrolase Activities of RT6 Alloantigens

The cloned vertebrate ADP-ribosyltransferases have significant amino acid sequence identity to each other and to the RT6 family of rat T cell alloantigens (Takada et al. 1995; Tsuchiya et al. 1994; Okazaki et al. 1995). In rats, *RT6^a* and *RT6^b* are alleles of a single gene locus that codes for the corresponding alloantigens RT6.1 and RT6.2, respectively. RT6.1 and RT6.2 differ in 5% of their sequences with ten, predominantly nonconservative, amino acid differences (Haag et al. 1990). Four clones obtained from a rat T cell hybridoma (EpD3) cDNA library demonstrated alternative splicing in the 5' region of the RT6.2 mRNA (Haag et al. 1993). A subset of peripheral lymphocytes from F_1(RT6[a] X RT6[b]) hybrid rats expressed only the RT6.2 alloantigen; the majority of lymphocytes expressed both RT6.1 and RT6.2 (Thiele et al. 1993). The RT6.2[+] cells did not express the RT6.1 mRNA consistent with pretranslational regulation of RT6 expression. The variable expression of RT6 alloantigens may be due to alternative usage of splice sites or alternative promoter usage (Thiele et al. 1993).

The expression of the RT6 proteins occurs late in maturation during T cell ontogeny and is restricted to peripheral T cells and intraepithelial lymphocytes (Thiele et al. 1987; Mojcik et al. 1988; Fangmann et al. 1990). The variably glycosylated RT6.1 and nonglycosylated RT6.2 are GPI-anchored proteins with molecular masses of 25–30-kDa (Thiele et al. 1986; Koch et al. 1988, 1990a).

The mouse homologue of the rat RT6 antigen has been cloned from spleen cells (Koch et al. 1990b). Unlike the rat, the mouse has two copies of the RT6 gene, *Rt6-1*, and *Rt6-2* (Prochazka et al. 1991). There are three distinct haplotypes, designated *Rt6[a]*, *Rt6[b]* and *Rt6[c]*, based on restriction fragment length polymorphisms (Koch-Nolte et al. 1995a). The *Rt6* and *H1* minor histocompatibility locus alleles on chromosome 7 are closely linked (Koch-Nolte et al. 1995a). The nucleic acid sequences of the two mouse *Rt6* genes are 85% identical, whereas the deduced amino acid sequences are 80% identical (Haag et al. 1994). The deduced amino acid sequence of the mouse *Rt6-1* protein is approximately 71% identical to that of the rat RT6.2 sequence (Koch et al. 1990b). The mouse *Rt6* genes are developmentally regulated, similar to that of the rat. In neonatal mice, *Rt6* expression is greater in lymphocytes from the intestine than in those from the spleen, whereas in adult animals, *Rt6* expression is greater in spleen than in intestinal lymphocytes. The two loci, however, are differentially expressed as determined by reverse transcription-polymerase chain reaction (RT-PCR) (Koch-Nolte et al. 1995b). The level of mRNA transcript from the *Rt6-1* locus is greater than that from *Rt6-2* in intestine and spleen of mice at all ages (Koch-Nolte et al. 1995b).

The RT6 gene structure in human and chimpanzee is highly conserved compared to the rodent species. The human RT6 gene, however, contains three premature stop codons corresponding to amino acids 47, 141, and 193 of the rat RT6 protein (Haag et al. 1994). In addition, no RT6 mRNA could be detected on Northern analysis of RNA from human tissues or by reverse transcriptase-polymerase chain reaction, procedures that readily detect RT6 RNA transcripts in rodent lymphoid tissue. The human single copy pseudogene has been localized to chromosome 11q13 (Koch-Nolte et al. 1993).

The absence of RT6$^+$ T cells in BB/Wor rats is associated with the occurrence of an autoimmune-mediated diabetes (Greiner et al. 1986; Burstein et al. 1989; Doukas and Mordes 1993). Diabetes can be prevented in diabetes-prone (DP BB/Wor) rats by transfusion and long-term engraftment of RT6$^+$ spleen cells (Rossini et al. 1986), but not thymocytes or bone marrow cells which contain few if any RT6$^+$ cells (Burstein et al. 1989). Depletion of RT6$^+$ cells from diabetes-resistant (DR BB/Wor) rats by injection of anti-RT6 monoclonal antibody, increased the incidence of diabetes in DR rats (Greiner et al. 1987). Prevention of diabetes in DP rats by spleen cell transfusion or the induction of diabetes in DR rats by RT6$^+$ cell depletion appeared to be an age-dependent phenomenon. The presence of RT6$^+$ T lymphocytes in $\leq$ 30–60 day old rats was crucial to prevent spontaneous diabetes in DR rats or to inhibit disease onset in DP rats. RT6$^+$ cell transfusion in DP rats after 60 days was ineffective in preventing onset of diabetes. Likewise, RT6$^+$ cell depletion after 60 days did not induce diabetes in DR rats (Burstein et al. 1989). In other experiments, DP rats in which engraftment of a pancreas transplant from DR rats was successful, demonstrated euglycemia, a reversal of lymphopenia, and a near normal level of splenic and peripheral RT6$^+$ T cells (Uchikoshi et al. 1995). DP rats with recurrent diabetes after transplatation, on the other hand, had a recurrence of hyperglycemia, no recovery of T cell count and no detectable splenic RT6$^+$ cells. The pancreatic grafts from these rats demonstrated a mononuclear cell infiltration of the islet cells.

The nonobese diabetic (NOD) mouse has reduced levels of *Rt6*-specific mRNA and is prone to develop autoimmune-mediated diabetes (Prochazka et al. 1987). The NZW mouse carries a spontaneous disruption of the *Rt6-2* gene, but the homozygous null mutation of *Rt6-2* is not associated with disease, despite low levels of *Rt6* transcript (Koch-Nolte et al. 1995b). The NZB mouse, on the other hand, has normal levels of *Rt6* but is prone to development of a variety of autoantibodies. Low *Rt6* mRNA levels alone, therefore, do not imply autoimmune disease susceptibility. (NZB X NZW)F₁ mice, like NZW mice, have shown reduced *Rt6* mRNA levels and manifest an autoimmune-mediated lupus-like glomerulonephritis.

NMU cells transformed with the rat RT6.2 cDNA exhibited GPI-linked, NAD glycohydrolase activity (Takada et al. 1994). RT6.1 and RT6.2, from EpSM30 and EpD3 lymphocytes, respectively, released with PI-PLC and immunoprecipitated, demonstrated NAD glycohydrolase activity (Haag et al. 1995). RT6.2, but not RT6.1, was capable of auto-ADP-ribosylation (Haag et al. 1995). Automodification of RT6.2 was observed in intact T cells, as well as with immunopurified RT6.2 (Haag et al. 1995). ADP-ribose itself failed to modify RT6.2. Auto-ADP-ribosylation of RT6.2 was reversed with 1 M NH$_2$OH, but not 10 mM HgCl$_2$, consistent with an ADP-ribose-arginine bond. RT6.2 contains two arginines that are lacking in RT6.1. These may serve as specific acceptor sites for ADP-ribosylation (Haag et al. 1995). Other investigators, however, demonstrated an arginine-specific auto-ADP-ribosylation of RT6.1 in intact peripheral T cells, but not from the membrane fraction (Maehama et al. 1995). The modified RT6.1, after release from the lymphocyte surface, was immunoprecipitated with anti-RT6.1 serum. Further, the ADP-ribosylation of RT6.1 was reversible, and this was inhibited by ADP-ribose, which is characteristic of a reaction catalyzed by an ADP-ribosylarginine hydrolase (Maehama et al. 1995). Whether RT6.1, RT6.2, or both are auto-ADP-ribosylated, may be influenced by the assay conditions. Crosslinking of RT6.2 on intact T cells with an anti-RT6.2 monoclonal antibody, enhanced auto-ADP-ribosylation of RT6.2 (Rigby et al. 1996). To determine whether the effect of NAD on RT6.2$^+$ T cells was similar to that described in mouse CTL (Wang et al. 1994), rat T cells were incubated with or without NAD, followed by evaluation of the proliferative response to T cell activation. Incorporation of [^{3}H]thymidine in response to three agents, anti-CD3, concanavalin A, or phorbol myristic acid (PMA) plus ionomycin was inhibited by 200 μM NAD, but not by 20 μM (Rigby et al. 1996). The inhibition of T cell and CTL proliferation by exogenous NAD or its metabolites could be mediated by the RT6 proteins, lymphocyte ADP-ribosyltransferases (Yac1 and Yac2), or other NAD-utilizing enzymes.

The mouse *Rt6* locus 1 gene product and rat RT6.2 proteins expressed in insect (Sf9) cells using a baculovirus expresssion system demonstrated auto-ADP-ribosyltransferase activity (Rigby et al. 1996). The recombinant mouse Rt6-1, but not the rat RT6.2, catalyzed the ADP-ribosylation of histones. In other experiments, recombinant mouse Rt6-1 and Rt6-2 proteins, affinity-purified using the Sepharose-bound M2-antibody, ADP-ribosylated the matrix-associated light chain of M2 (Koch-Nolte et al. 1996). All of the resulting ADP-ribose-protein bonds were NH$_2$OH-labile, consistent with arginine-specific ADP-ribosyltransferase activity (Rigby et al. 1996).

2.4
Inhibitors of Mono-ADP-Ribosyltransferases

In general, three enzymes, poly(ADP-ribose) polymerase, mono-ADP-ribosyltransferase, and NAD glycohydrolase, appear to be affected by a common pool of inhibitors (Banasik and Ueda 1994). Some compounds are more selective in their inhibition of one or more of these enzymes. Inhibitors of mono-ADP-ribosyltransferase and poly(ADP-ribose) polymerase activities are shown in Table 1. The effective concentration of some of the inhibitors that preferentially block poly(ADP-ribose) polymerase activity, is two or more orders of magnitude lower than that required to block mono-ADP-ribosyltransferase activity. For example, the concentration that inhibited polymerase activity by 50% (IC_{50}) for benzamide and its derivatives was 3–6 μM, whereas the IC_{50} for turkey transferase A was

Table 1. 50% inhibitory concentration (IC_{50}) of compounds that inhibit mono-ADP-ribosyltransferases and poly(ADP-ribose) polymerase (from Rankin et al. 1989; Banasik et al. 1992)

Compound	IC_{50} Transferase (μM)	IC_{50} Polymerase (μM)
5'Bromodeoxyuridine	590 ± 94	15 ± 1.3
Thymidine	1900 ± 300	43 ± 5.2
3-Methoxybenzamide	2700 ± 250	3.4 ± 0.31
Theophylline	2800 ± 460	46 ± 15
3-Aminobenzamide	3000 ± 970	5.4 ± 0.40
Nicotinamide	3400 ± 410	43 ± 5.2
Benzamide	4100 ± 220	3.3 ± 0.28
Arachidic acid (C20:0)	4.4	>1000
Stearic acid (C18:0)	6.1	>1000
Palmitic acid (C16:0)	16	>1000
Arachidonic acid (C20:4)	66	44
Linoleic acid (C18:2)	90	48
Linolenic acid (C18:3)	110	110
Palmitoleic acid (C16:1)	200	95
Vitamin K_1	1.9	520
Vitamin $K_{2(20)}$	13	ND[a]
Novobiocin	280	2200

[a]ND, not determined

2700–4100 μM (Rankin et al. 1989). Similarly, 5'-bromodeoxyuridine had IC$_{50}$ values of 15 and 590 μM for the polymerase and transferase A, respectively. These differences may be useful in differentiating pathways involving poly(ADP-ribosyl)ation from those dependent on mono(ADP-ribosyl)ation.

Vitamins K$_1$ (phylloquinone) and K$_{2(20)}$ (menaquinone), and saturated long-chain fatty acids arachidic, stearic, and palmitic acids were specific inhibitors of mono-ADP-ribosyltransferase activity with IC$_{50}$ values between 1.9 and 16 μM (Banasik et al. 1992). The unsaturated long-chain fatty acids, on the other hand, did not differentially inhibit the heterophil transferase and the poly(ADP-ribose)polymerase. Novobiocin (Table 1) and the flavanoid sylbin (Donadoni et al. 1995) are two other natural compounds that specifically inhibit the transferase. These studies were conducted with a limited number of ADP-ribosylating enzymes, and effects of inhibitors on other members of these families may well differ (e.g., IC$_{50}$, specificity).

3
Regions of Conserved Structure Among ADP-Ribosyltransferases

Several bacterial toxin ADP-ribosyltransferases appear to have three non-contiguous regions of similarity despite an overall lack of amino acid sequence identity (Rappuoli and Pizza 1991; Domenighini et al. 1994). Region I contains a nucleophilic arginine or histidine, which is possibly involved in hydrogen bonding. Region II, which is composed of closely spaced aromatic and hydrophobic amino acids, which form a pocket for binding of the nicotinamide moiety and adenine ring of NAD, is located approximately 50–75 amino acids downstream of region I. Region III, located approximately 100–150 amino acids following region II, contains a glutamate that serves as the catalytic-site residue for NAD hydrolysis and ADP-ribose transfer. The crystal structures of DT (Choe et al. 1992; Bennett and Eisenberg 1994), LT (Sixma et al. 1991), ETA (Allured et al. 1986) and PT (Stein et al. 1994) revealed catalytic clefts formed, in part, by the three conserved regions. When the toxins were photolabeled with NAD, the nicotinamide moiety was crosslinked to the active site glutamate (Carroll et al. 1985). The vertebrate transferases and the RT6.2 enzymes also share regions of amino acid sequence similarity to the bacterial toxin transferases, which is consistent with the view that many of the mono-ADP-ribosyltransferases have similar NAD-binding and catalytic sites (Takada et al. 1995).

3.1
Region I: Participation of Arginine and Histidine Residues in Hydrogen Bonding

His21 of DT, His440 of ETA, Arg7 of CT and the structurally and immunologically related LT, and Arg9 of PT are the nucleophilic amino acids responsible for NAD binding (Fig. 1a). Conversion of His21 to N-carbethoxyhistidine following incubation of DT with diethylpyrocarbonate (DEPC) resulted in the inhibition of NAD binding and ADP-ribosyltransferase and NAD glycohydrolase activities (Papini et al. 1989). Replacement of His21 with other amino acids, except asparagine, drastically reduced ADP-ribosyltransferase activity (Johnson and Nicholls 1994; Blanke et al. 1994a). The asparagine substitution caused only modest reductions in K_m value for NAD, catalytic rate (k_{cat}), and catalytic efficiency (k_{cat}/K_m). The K_d value for NAD was only ten times that of the wild-type toxin (measured by quenching of intrinsic protein fluorescence) compared to a more than 30-fold increase for the other mutants. A marked reduction in NAD glycohydrolase activity was observed with all mutations, including the asparagine substitution (Blanke et al. 1994a). Replacement of His21 in DT with the sterically similar asparagine without abolishing NAD-binding or enzyme activity is consistent with the suggestion that the π-nitrogen of the imidazole group of histidine is important in orienting NAD in the catalytic site by hydrogen bonding to the carboxamide group of nicotinamide (Johnson and Nicholls 1994; Blanke et al. 1994a).

Based on X-ray crystallography, the position of His440 of ETA corresponds to that of His21 of DT (Carroll and Collier 1988). Replacement of His440 with alanine, phenylalanine or asparagine reduced ADP-ribosyltransferase activity more than 1000-fold and reduced the ability of the toxin to inhibit EF-2 (Han and Galloway 1995). NAD glycohydrolase activity, however, was not inhibited by these replacements; the asparagine replacement resulted in a tenfold increase in NADase activity (Han and Galloway 1995). NAD binding was unaffected, as the K_m values for NAD of the mutated toxins were similar to that of the wild-type ETA. Replacement in ETA of His426, positioned on the external surface of the catalytic cleft (Allured et al. 1986) with tyrosine, abolished transferase activity without altering NAD binding (Wozniak et al. 1988). His426 (Kessler and Galloway

Region I

```
                    *
RMT     177 VFRGVHG
Yac1    172 ·y·····
Yac2    159 ·····g·
Rt6-1   144 ·y··TNV
RT6.1   144 ·y··Tkt
RT6.2   144 ·y··Tkt
AT1     162 ·y···r·
AT2     162 ·y···r·
DT       19 Syh·Tkp
ETA     438 Gyh·TFL
LT        5 ly·aDSR
```

Region II

```
DT       48 DDW-K-GF-YSTDNKYDAAGY
ETA     464 AI·-r-····IAGdPAL·Y··
LT       79 GYS--TYyI·VIATAPnM--f
```

Region III

```
                    *
RMT     231 GYSFFPGEEEVLIP
Yac1    226 ······E·······
Yac2    213 aL·V··E·R·····
Rt6-1   200 HC·yytH·······
RT6.1   200 Ef··y·Dq······
RT6.2   200 Ef··R·Dq······
AT1     215 Qf····s·d·····
AT2     215 Qf··y·s·d·····
DT      136 EG·SSV---·YiNN
ETA     544 ·PEEEg·RL·Tilg
LT      105 ---·PH·Y·q··SAL
C3      164 pI·y···qL···l·
T2      581 ·SLGIat·A··il·
T4      568 ·SLApsN·M··il·
T6      581 ·SLGIat·A··il·
```

1992) and His440 (Han and Galloway 1995) of ETA appear to be involved in the transfer of ADP-ribose to EF-2 or directly in the interaction with EF-2, and not in hydrogen bond formation with NAD as was hypothesized for His21 of DT (Han and Galloway 1995).

Arg7 of LT and CT was found to be essential for toxin activity based on crystal structure and site-directed mutagenesis. Substituting lysine for Arg7 in LT (Lobet et al. 1991) and CT (Burnette et al. 1991) resulted in loss of enzymatic activity, and increased the sensitivity of LT to trypsin digestion. As determined by computer modeling, Arg7 participates with Val53 and Arg54 in a hydrogen bond network, which maintains the required conformation at the enzymatic site of LT. Although lysine has the same charge and size as arginine, it could not form the necessary hydrogen bonds at the active site to support catalysis (Pizza et al. 1994). Crystal structure of LT demonstrated that Arg7 also forms a hydrogen bond with Ser61 in the active site (Sixma et al. 1993). Replacement of Ser61 with threonine reduced by only a factor of 10, the k_{cat} of the ADP-ribosyltransferase reaction of LT without affecting the K_m values for NAD and the ADP-ribose acceptor agmatine, suggesting that Ser61 functions to maintain proper conformation of the active site and does not play an essential role in catalysis (Cieplak et al. 1995). Similarly, the crystal structure of PT revealed that the side chain of Arg9 projected into the active site, where it

---

Fig. 1 a–c. Regions of sequence similarity among vertebrate and bacterial toxin ADP-ribosyltransferases are aligned with seemingly analogous regions in the rabbit muscle (region I and III) or diphteria toxin (region II) transferase. a Region I containing a nucleophilic histidine or arginine (*asterisk*). b Region II containing closely spaced aromatic and hydrophobic amino acids (*underlined*). c Region III with alignment of active-site glutamic acid (*asterisk*). Sequences are in the single letter code with the position number of the first amino acid following the name of the protein. • indicates amino acid identical to that in rabbit muscle (a and c) or diptheria toxin (b) transferase. Lower case letter indicates conservative difference from the rabbit transferase. – indicates a gap to optimize alignment. *RMT*, rabbit muscle transferase; *Yac1*, mouse lymphocyte ADP-ribosyltransferase Yac1; *Yac2*, mouse lymphocyte ADP-ribosyltransferase Yac2; *RT6.1*, rat T cell alloantigen RT6.1; *RT6.2*, rat T cell alloantigen RT6.2; *Rt6-1*, mouse T cell alloantigen *Rt6* locus 1 protein; *AT1*, chicken ADP-ribosyltransferase 1; *AT2*, chicken ADP-ribosyltransferase 2; *DT*, diphtheria toxin; *ETA*, *Pseudomonas aeruginosa* exotoxin A; *LT*, heat-labile enterotoxin of *Escherichia coli*; *C3*, *Clostridium botulinum* C3 exotoxin; *T2*, ADP-ribosyltransferase (gene product Alt) of bacteriophage T2; *T4*, ADP-ribosyltransferase of bacteriophage T4; *T6*, ADP-ribosyltransferase of bacteriophage T6

formed a hydrogen bond with Ser52, holding together adjacent strands of the catalytic cleft (Stein et al. 1994). An Arg9 to lysine mutation in PT (Burnette et al. 1988; Kaslow et al. 1989) resulted in complete loss of activity.

Alignment of deduced amino acid sequences of the rabbit muscle transferase with those of several of the bacterial toxin transferases along with data from computer modeling of the mouse Rt6 proteins (Koch-Nolte et al. 1996) demonstrated that Arg179 of the muscle transferase and Arg126 of the mouse Rt6-1 and Rt6-2 transferases are the critical H-R region residues. Similarly, Arg174 and Arg161 of the Yac1 and Yag2 transferases, respectively, and Arg164 of the chicken heterophil enzymes, AT1 and AT2, align with the critical arginines of the muscle transferase and mouse Rt6 enzymes.

3.2
Region II: Involvement of an Aromatic- and Hydrophobic-rich Domain in Substrate Binding

Conservation of aromatic amino acids hypothesized to be involved in hydrophobic interactions with the aromatic rings of NAD was evident in the alignment of amino acids in the active site cleft of DT and ETA; Trp50, Phe53, Tyr54 and Tyr65 of DT correspond to Trp466, Phe469, Tyr470 and Tyr481 of ETA (Carroll and Collier 1988). Photolabeling of DT with 8-azidoadenine and 8-azidoadenosine suggested stacking of the nicotinamide moiety of NAD on the phenolic ring of Tyr65 and the adenine ring on Trp50 (Papini et al. 1991). Crystal structure of DT complexed to adenylyl-3',5'-uridine monophosphate (AppU), a dinucleotide structurally similar to NAD, positioned the nicotinamide ring adjacent to the imidazole ring of His21, the phenolic ring of Tyr65 and the side chain of the active site Glu148 (Choe et al. 1992). Replacement of Trp50 with phenylalanine in DT resulted in minimal reduction of NAD-binding and ADP-ribosyltransferase activity, whereas replacement with alanine (T50A), eliminated NAD glycohydrolase activity, markedly reduced (5,000-fold) transferase activity (k_{cat}), and decreased efficiency 6×10^4-fold (k_{cat}/K_m) (Wilson et al. 1994). Although the K_d for NAD binding by the Thr50Ala mutant was too great to be measured by fluorescence quenching, the K_m for NAD in the transferase reaction with EF-2 as ADP-ribose acceptor was only approximately ten times that for wild-type DT. It appears that EF-2 stabilized NAD in the active site pocket during ADP-ribosylation, an effect absent during NAD hydrolysis (Wilson et al. 1994).

Whereas replacement of Thr65 of DT with phenylalanine (Thr65Phe) reduced ADP-ribosyltransferase activity approximately fourfold without altering NAD binding, an alanine replacement (Thr65Ala), decreased NAD binding and reduced transferase activity 350-fold (Blanke et al. 1994b). The Thr65Phe mutant bound adenosine with an affinity equal to that of wild-type DT. Thr65Ala, however, had a slightly higher affinity for adenosine suggesting that nicotinamide interacted with Tyr65 and not with adenine or the adenine-ribose moieties of NAD (Blanke et al. 1994b). Although Trp153 is also located in the active site of DT, substitution of Ala for Trp153 had only moderate effects on NAD glycohydrolase and ADP-ribosyltransferase activities (Wilson et al. 1994). In the proposed model for NAD binding to DT, the adenine ring of NAD is adjacent to the indole ring of Trp50, and the nicotinamide ring is stacked against the phenolic ring of Tyr65 where it interacts with the imidazole ring of His21 via a hydrogen bond. These interactions limit the rotational and translational freedom of the N-glycosidic bond, providing a favorable orientation for nucleophilic attack in ADP-ribose transfer or NAD glycohydrolase reactions (Domenighini et al. 1991).

The aromatic amino acid-rich segment in PT encompasses amino acids 82–98, based on sequence and structural alignment of toxins (Rappuoli and Pizza 1991). The active site Tyr65 of DT is aligned with Phe97 of PT and Phe95 of CT and LT (Blanke et al. 1994b).

3.3
Region III: Characterization of a Critical Active-Site Glutamate

Photoaffinity labeling, site-directed mutagenesis and three-dimensional studies, demonstrate an active-site glutamic acid strictly conserved among several bacterial toxins (Domenighini et al. 1994). Photoaffinity labeling of DT resulted in lysis of the nicotinamide-ADP-ribose bond of NAD, decarboxylation of Glu148, and formation of a new bond between the γ-methylene carbon of Glu148 and carbon-6 of the nicotinamide ring (Carroll et al. 1985). Further, replacement of Glu148 with aspartate inhibited ADP-ribosylation of EF-2 by DT without affecting NAD binding (Tweten et al. 1985). Glu553 of ETA was found analogous to Glu148 of DT by photoaffinity labeling (Carroll and Collier 1987) and site-directed mutagenesis (Douglas and Collier 1987); similar properties were demonstrated for Glu129 of PT (Barbieri et al. 1989; Pizza et al. 1988). Replacement of Glu129 with aspartate inhibited the NAD glycohydrolase activity of PT, with minimal effects on K_m or K_d for NAD (Antoine et al. 1993).

His35 of PT resides in the catalytic site within hydrogen-bonding distance of Glu129 (Stein et al. 1994). Mutation of His35 to asparagine drastically reduced ADP-ribosyltransferase activity (Kaslow et al. 1989). Glutamine substitution for His35 had no effect on NAD or ADP-ribose acceptor (transducin) binding, but did reduce both transferase and NAD glycohydrolase activities by decreasing catalytic rates of each reaction (Xu et al. 1994). Replacement of His35 with proline, on the other hand, totally abolished enzyme activity without affecting NAD binding. Since glutamine, but not proline or asparagine, can mimic the hydrogen bonding capacity of the ε-N of the imidazole group, His35 appeared to enhance ADP-ribosylation by hydrogen bond formation between the acceptor protein (cysteine), or water molecule in the hydrolytic reaction. The ε-N, and the activated cysteine or water molecule would then function as a nucleophile to attack the N-glycosidic linkage of NAD (Xu et al. 1994; Antoine and Locht 1994). The carboxylate group of Glu129 in PT is postulated to stabilize a transition state intermediate involving the anomeric carbon of NAD, the nicotinamide and the incoming substrate prior to ADP-ribose transfer (Antoine and Locht 1994; Domenighini et al. 1994).

The enzymatic domain containing the active site glutamate of LT or CT (Glu112) and ETA (Glu553) could be structurally superimposed (Sixma et al. 1991) and resembled the glutamic acid residues in the active sites of DT and PT (Pizza et al. 1994). Mutagenesis of Glu110 or Glu112 of LT dramatically reduced ADP-ribosyltransferase activity (Tsuji et al. 1990; Lobet et al. 1991; Pizza et al. 1994). A mutant LT with a lysine substituted for Glu112 was without enzymatic and biologic activity, but remained immunologically identical to wild-type LT (Tsuji et al. 1990) and capable of interacting with its allosteric activator, ARF (Moss et al. 1993). Replacement of Glu112 with aspartate reduced by a factor of 100, the k_{cat} for ADP-ribosyltransferase activity using agmatine as the acceptor molecule; there was little effect, however, on the K_m values for NAD and agmatine (Cieplak et al. 1995). In contrast, the k_{cat} for transferase activity of LT was reduced by a factor of approximately 20 in a Gly110Asp mutant, consistent with the hypothesis that Glu112 may be more critical than Glu110 in the ADP-ribose transfer reaction (Cieplak et al. 1995). Replacement of Val97 of LT, which is buried within the toxin, with lysine resulted in the loss of ADP-ribosyltransferase activity (Merritt et al. 1995). The crystal structure of the mutant LT revealed that the volume of a H_2O-filled cavity is reduced

64% by the replacement of the valine side chain by the larger lysine. Further, the charged amino group of the lysine forms a salt bridge with the carboxylate of Glu112, which may alter the catalytic activity of the essential glutamate. Alternatively, reduction in size of the H_2O-filled cavity may limit the ability of the toxin to undergo a conformational change necessary for the active-site cleft to accept substrates of the transferase reaction (Merritt et al. 1995).

Other bacterial toxin ADP-ribosyltransferases possess active site glutamates. *Clostridium limosum* C3-like toxin, an ADP-ribosyltransferase that modifies asparagine 41 of the Rho family of GTP-binding proteins (Sekine et al. 1989; Braun et al. 1989; Narumiya et al. 1988), contains a glutamate that was photolabeled with NAD and corresponds to Glu173 of the related *C. botulinum* exoenzyme C3 (Nemoto et al. 1991; Jung et al. 1993). Replacement of Glu174 of *C. limosum* with aspartate or glutamine reduced ADP-ribosyltransferase activity by more than 1000-fold, but only minimally reduced NAD binding (Bohmer et al. 1996). An active site glutamate was also photolabeled in *Bacillus cereus* exoenzyme, another, structurally distinct, toxin that ADP-ribosylates Rho (Just et al. 1995b). Exoenzyme S from *P. aeruginosa* (ExoS), cloned from strain 388 (Kulich et al. 1994), ADP-ribosylates a number of proteins in vitro in the presence of a eukaryotic accessory protein, termed factor-activating exoenzyme S or FAS (Coburn et al. 1991); ExoS has been implicated as a virulence factor of *P. aeruginosa* in burns and chronic lung infections. The FAS-dependent ADP-ribosyltransferase activity was localized to the carboxy-terminal 222 amino acids (Knight et al. 1995). The BESTFIT algorithim revealed a high degree of identity between amino acids 333 and 352 of ExoS and similar regions of LT, CT, *C. botulinum* exoenzyme C3, and aligned Glu265 of ExoS with Glu112 of LT and CT (Kulich et al. 1994).

Of the three glutamic acids at positions 238–240 of the skeletal muscle ADP-ribosyltransferase (Fig. 1c), Glu240 and the surrounding amino acids were aligned with the region containing the catalytic glutamate of several bacterial toxins (Takada et al. 1995). Replacement of Glu240 with aspartate or alanine resulted in loss of transferase activity, whereas Glu239Asp retained activity. Substitution of aspartate or glutamine for Glu238, on the other hand, yielded inactive proteins. Glu238 and Glu240 of the muscle transferase were postulated to be functionally analogous to Glu110 and Glu112 of LT and CT (Takada et al. 1995).

The Glu278-Tyr-Ile sequence of the muscle transferase, similar to that found at the active site of DT (Tweten et al. 1985) and poly(ADP-ribose) synthetase (Marsischky et al. 1992), is another potential active-site glutamate. Substitution of aspartate or alanine for Glu278, however, did not affect transferase activity. Likewise, Glu282 of the muscle enzyme was aligned with Asp993 of poly(ADP-ribose) synthetase, another critical catalytic site residue (Marsischky et al. 1992). Replacement of Asp993 with alanine (but not glutamate), inactivated poly(ADP-ribose) synthetase (Simonin et al. 1993), however, whereas, alanine or aspartate substitution for Glu282 in the muscle transferase, did not affect its activity (Takada et al. 1995). Based on sequence similarity and site-specific mutations, Glu240 in the muscle transferase appears to correspond to the critical glutamate of the bacterial toxins.

The NAD glycohydrolase, RT6.2, contains a similar cluster of acidic amino acids (Gln207, Glu208 and Glu209), which can be aligned with glutamates 238–240 of the muscle transferase. Moreover, the region surrounding Glu222, Asp223 and Glu224 of the chicken heterophil transferases, AT1 and AT2, has significant similarity to the active site region of the muscle transferase (Tsuchiya et al. 1994). In the ADP-ribosyltransferases of bacteriophages T2, T4 and T6 (gene product Alt), which ADP-ribosylate and inhibit the host *E. coli* RNA polymerase α-subunit (Koch and Ruger 1994), a similar region is found, although a methionine or alanine is present between the two glutamates. Based on mutagenesis studies with the rabbit transferase, the presence of a nonacidic residue amino terminal to the critical glutamate should not affect activity (Takada et al. 1995).

The soluble mouse lymphocyte antigen CD38 (sCD38) contains glutamate and aspartate residues at positions 150 and 151 (146 and 147 in the human sequence) that may serve as active site residues (Grimaldi et al. 1995). The simultaneous replacement of Glu150 with leucine and Asp151 with valine completely abolished NAD glycohydrolase activity and prevented ADP-ribosylation of lysozyme without altering sCD38 structure. Other substitutions demonstrated that Glu150 appears to be more important than Asp151 for enzymatic activity (Grimaldi et al. 1995).

4
ADP-Ribosylarginine Hydrolases

Mono-ADP-ribosyltransferases and ADP-ribosylarginine hydrolases cata-
lyze the forward and reverse reactions, respectively, of a putative ADP-ribosylation cycle. Hydrolases have been identified in mammalian, avian
and bacterial systems and appear to be ubiquitous in eukaryotic tissues.

4.1
Turkey Erythrocyte ADP-Ribosylarginine Hydrolase

The turkey erythrocyte hydrolase, is a soluble, 39-kDa monomeric protein
(Moss et al. 1988). The hydrolase was maximally activated by 5–10 mM
Mg^{2+} plus 5–10 mM dithiothreitol and was inhibited more than 80% by
5 mM NaF or 200 mM NaCl; histones had no effect on activity (Moss et
al. 1986).

The ADP-ribose, but not the arginine moiety, was critical for substrate
recognition and degradation by the hydrolase (Moss et al. 1986). The hydrolase was competitively inhibited by ADP-ribose > ADP > AMP; arginine, agmatine and guanidine had no effect. The hydrolase cleaved ADP-
ribosylarginine, (2'phospho-ADP-ribosyl)arginine, and ADP-ribosylguanidine. (Phosphoribosyl)arginine and ribosylarginine, products of
sequential degradation of ADP-ribosylarginine by phosphodiesterase and
phosphatase, were neither substrates nor inhibitors of the hydrolase. The
K_m for ADP-ribosylarginine was 65 μM; those for (2'phospho-ADP-ribosyl)arginine and ADP-ribosylguanidine were 47 μM and 27 μM, respectively. Arginine, produced by hydrolase-catalyzed degradation of ADP-ribosylarginine, served as substrate for the purified turkey ADP-ribosyltransferase (Moss et al. 1985), demonstrating preservation of the guanidino
moiety during hydrolysis. Moreover, the stereospecificity of the hydrolase
matched that of the turkey ADP-ribosyltransferase. In the presence of
β-NAD and arginine, the transferase catalyzed the synthesis of the α-
anomeric ADP-ribosylarginine whereas the hydrolase utilized α-ADP-ri-
bosylarginine, but not the β-anomer, as substrate (Moss et al. 1986). This
finding is compatible with the hypothesis that the ADP-ribosyltransferases
and ADP-ribosylarginine hydrolases act as opposing arms of an ADP-ribosylation cycle.

The hydrolase was inactivated after ADP-ribosylation by the turkey
transferase. Inactivation was prevented by the presence of Mg^{2+} or Mg^{2+}

plus dithiothreitol. The activity of the ADP-ribosylated hydrolase could not be restored, however, by addition Mg^{2+} or Mg^{2+} and dithiothreitol, nor did these agents alter ADP-ribosyltransferase activity. The hydrolase appeared to be "stabilized" by Mg^{2+} and dithiothreitol in a conformation resistant to transferase-catalyzed ADP-ribosylation of critical arginine residues (Moss et al. 1988).

4.2
Mammalian ADP-Ribosylarginine Hydrolases

An ADP-ribosylarginine hydrolase was purified approximately 20, 000-fold from rat brain. Hydrolases from other rat tissues and mouse, guinea pig, rabbit, sheep, pig and calf brain were partially purified for structural and functional comparison with that from rat brain (Moss et al. 1992). Among rat tissues, hydrolase activity was highest in brain, spleen and testis. Hydrolase activities from rat and mouse brain were higher than those from the other mammalian species tested. Similar to the turkey hydrolase, the rat and mouse brain enzymes were synergistically stimulated by Mg^{2+} and dithiothreitol. In contrast, the pig and calf hydrolases were stimulated by Mg^{2+} but not dithiothreitol. Polyclonal rabbit anti-rat brain hydrolase antibodies reacted on immunoblot with 39-kDa proteins from turkey erythrocytes and mouse, rat and calf brains despite differences in enzymatic activity.

A cDNA probe based on amino acid sequence of a tryptic peptide from the purified rat hydrolase was used to isolate from a rat brain cDNA library, a hydrolase coding region cDNA containing an open reading frame of 1089 bp (Moss et al. 1992). The cloned hydrolase cDNA, expressed as a glutathione S-transferase-linked fusion protein in *E. coli*, had Mg^{2+}- and dithiothreitol-dependent hydrolase activity and reacted on immunoblot with anti-hydrolase antibodies. A hydrolase-specific oligonucleotide probe hybridized on Northern blot, with a 1.7-kb mRNA in total RNA from all rat tissues. Levels were highest in brain, spleen, testis, and lung and correlated with amounts of enzyme activity. A PCR-generated hydrolase cDNA probe detected a 1.7-kb band in poly $(A)^+$ RNA from rat and mouse brain, but not from chicken, rabbit, or bovine brain, or cultured IMR-32 or HL-60 cells (Moss et al. 1992).

Mouse and human hydrolase genes were cloned by PCR using oligonucleotide primers generated from the rat brain hydrolase cDNA (Takada et al. 1993). The mouse hydrolase was 92% and 94% identical in nucleotide and deduced amino acid sequences, respectively, to those of the rat. Nucleotide and deduced amino acid sequences of the human and rat hydrolases were 82% and 83% identical, respectively. On Northern blot, human hydrolase-specific oligonucleotide probes hybridized with a 4-kb band using poly(A)$^+$ RNA from human brain, lung, and placenta, but not from IMR-32 or undifferentiated HL-60 cells.

In the rat and mouse hydrolases, the positions of five cysteines were identical whereas only four of the five cysteines were present in the human hydrolase (Takada et al. 1993). Cysteine 108 in the rat and mouse enzymes was replaced by a serine at position 103 in the human sequence. Based on the fact that the hydrolase in rat tissues was activated by dithiothreitol but activity in human tissues was thiol-independent, site-directed mutagenesis of the rat and human hydrolases was utilized to determine whether this difference was related to the fifth cysteine (Takada et al. 1993). Expression in E. coli of a mutant human hydrolase in which cysteine was substituted for serine at position 103 (Ser103Cys) produced a thiol-dependent enzyme similar to the native rat hydrolase. Wild-type human hydrolase expressed in E. coli retained its thiol-independence. There were no differences between wild-type and mutant enzymes in Mg^{2+} requirement, specific activity, or K$_m$ for ADP-ribosylarginine. On the other hand, the mutated rat hydrolase containing a serine at position 108 instead of cysteine (Cys108Ser), demonstrated thiol-independence, in contrast to the similarly expressed thiol-dependent, wild-type rat hydrolase. Both recombinant rat proteins had Mg^{2+} requirements and specific activities similar to those of the native rat hydrolase.

Anti-hydrolase antibodies raised against the rat brain hydrolase reacted on immunoblot with the native and recombinant wild-type rat hydrolases, but only weakly with the mutant Cys108Ser rat enzyme and not at all with the wild-type (Ser103) or mutant (Ser103Cys) human hydrolases. The differences observed in enzymatic properties as well as immunoreactivity may be due to single amino acid differences among native hydrolases (Takada et al. 1993).

The rat ADP-ribosylarginine hydrolase, expressed as a fusion protein in E. coli, released the ADP-ribose moiety from G$_{s\alpha}$ that had been ADP-ribosylated by CT, and from the auto-ADP-ribosylated A$_1$ subunit of CT (Maehama et al. 1994). Nonmuscle actin ADP-ribosylated by botulinum C2 toxin (Vandekerckhove et al. 1988) also served as a substrate for the hydrolase. EF-2 ADP-ribosylated by DT, G$_{o\alpha}$ modified by PT, and the GTP-binding protein Rho modified by C3 exoenzyme, however, were not affected by the recombinant hydrolase (Maehama et al. 1994), consistent with the specificity of the hydrolase for the ADP-ribose-arginine bond.

5
Summary

ADP-ribosylation of proteins has been observed in numerous animal tissues including chicken heterophils, rat brain, human platelets, and mouse skeletal muscle. ADP-ribosylation in these tissues is thought to modulate critical cellular functions such as muscle cell development, actin polymerization, and cytotoxic T lymphocyte proliferation. Specific substrates of the ADP-ribosyltransferases have been identified; the skeletal muscle transferase ADP-ribosylates integrin $\alpha 7$ whereas the chicken heterophil enzyme modifies the heterophil granule protein p33 and the CTL enzyme ADP-ribosylates the membrane-associated protein p40. Transferase sequence has been determined which should assist in elucidating the role of ADP-ribosylation in cells.

There is sequence similarity among the vertebrate transferases and the rodent RT6 alloantigens. The RT6 family of proteins are NAD glycohydrolases that have been shown to possess auto-ADP-ribosyltransferase activity whereas the mouse Rt6-1 is also capable of ADP-ribosylating histone. Absence of RT6$^+$ T cells has been associated with the development of an autoimmune-mediated diabetes in rodents. Humans have an RT6 pseudogene and do not express RT6 proteins.

The reversal of ADP-ribosylation is catalyzed by ADP-ribosylarginine hydrolases, which have been purified and cloned from rodent and human tissues. In principle, the transferases and hydrolases could form an intracellular ADP-ribosylation regulatory cycle. In skeletal muscle and lymphocytes, however, the transferases and their substrates are extracellular membrane proteins whereas the hydrolases described thus far are cytoplasmic. In cultured mouse skeletal muscle cells, processing of the ADP-ribosylated integrin $\alpha 7$ was carried out by phosphodiesterases and possibly phosphatases, leaving a residual ribose attached to the (arginine)protein.

Several bacterial toxin and eukaryotic mono-ADP-ribosyltransferases, and perhaps other NAD-utilizing enzymes such as the RT6 alloantigens share regions of amino acid sequence similarity, which form, in part, the catalytic site. The catalytic cleft, found in the bacterial toxins that have been studied thus far, contains a critical glutamate and other amino acids that function to position NAD for nucleophilic attack at the N-glycosidic linkage, for either ADP-ribose transfer or NAD hydrolysis. Amino acid differences among the transferases at the active site may be required for accommodating the different ADP-ribose acceptor molecules.

Acknowledgment. We thank Dr. Martha Vaughan for helpful discussions and critical review of the manuscript and Carol Kosh for expert secretarial assistance.

References

Abbracchio MP, Cattabeni F, Di Giulio AM, Finco C, Paoletti AM, Tenconi B, Gorio A (1991) Early alteration of G_i/G_o protein-dependent transductional processes in the retina of diabetic animals. J Neurosci Res 29:196–200

Aktories K (1994) Clostridial ADP-ribosylating toxins: effects on ATP and GTP-binding proteins. Mol Cell Biochem 138:167–176

Aktories K, Barmann M, Ohishi I, Tsuyama S, Jakobs KG, Habermann E (1986) Botulinum C2 toxin ADP-ribosylates actin. Nature 322:390–392

Allured VS, Collier RJ, Carroll SF, McKay DB (1986) Structure of exotoxin A of *Pseudomonas aeruginosa* at 3.0-Angstrom resolution. Proc Natl Acad Sci USA 83:1320–1324

Alvarez-Gonzales R, Pacheco-Rodriguez G, Mendoza-Alvarez H (1994) Enzymology of ADP-ribose polymer synthesis. Mol Cell Biochem 138:33–57

Antoine R, Locht C (1994) The NAD-glycohydrolase activity of the pertussis toxin S1 subunit: involvement of the catalytic His-35 residue. J Biol Chem 269:6450–6457

Antoine R, Tallett A, van Heyningen S, Locht C (1993) Evidence for a catalytic role of glutamic acid 129 in the NAD-glycohydrolase activity of the pertussis toxin S1 subunit. J Biol Chem 268:24149–24155

Banasik M, Ueda K (1994) Inhibitors and activators of ADP-ribosylation reactions. Mol Cell Biochem 138:185–197

Banasik M, Komura H, Shimoyama M, Ueda K (1992) Specific inhibitors of poly(ADP-ribose) synthetase and mono(ADP-ribosyl)transferase. J Biol Chem 267:1569–1575

Barbieri JT, Mende-Mueller LM, Rappuoli R, Collier RJ (1989) Photolabeling of Glu-129 of the S1 subunit of pertussis toxin with NAD. Infect Immun 57:3549–3554

Bennett MJ, Eisenberg D (1995) Refined structure of monomeric diphtheria toxin at 2.3 Angstrom resolution. Protein Sci 3:1464–1475

Blanke SR, Huang K, Wilson BA, Papini E, Covacci A, Collier RJ (1994a) Active-site mutations of diphtheria toxin catalytic domain: role of histidine-21 in nicotinamide adenine dinucleotide binding and ADP-ribosylation of elongation factor 2. Biochemistry 33:5155–5161

Blanke SR, Huang K, Collier RJ (1994b) Active-site mutations of diphtheria toxin: role of tyrosine-65 in NAD binding and ADP-ribosylation. Biochem 33:15494–15500

Bohmer J, Jung M, Sehr P, Fritz G, Popoff M, Just I, Aktories K (1996) Active site mutation of the C3-like ADP-ribosyltransferase from *Clostridium limosum* – analysis of glutamic acid 174. Biochemistry 35:282–289

Braun U, Habermann B, Just I, Aktories K, Vandekerckhove J (1989) Purification of the 22-kDa protein substrate of botulinum ADP-ribosyltransferase C3 from porcine brain cytosol and its characterization as a GTP-binding protein highly homologous to the *rho* gene product. FEBS Lett 243:70–76

Burnette WN, Cieplak W, Mar VL, Kaljot KT, Sato H, Keith JM (1988) Pertussis toxin S1 mutant with reduced enzyme activity and a conserved protective epitope. Science 242:72–74

Burnette WN, Mar VL, Platler BW, Schlotterbeck JD, McGinley MD, Stoney KS, Rohde MF, Kaslow HR (1991) Site-specific mutagenesis of the catalytic subunit of cholera toxin: substitution lysine for arginine 7 causes loss of activity. Infect Immun 59:4266–4270

Burstein D, Mordes JP, Greiner DL, Stein D, Nakamura N, Handler ES, Rossini AA (1989) Prevention of diabetes in BB/Wor rat by single transfusion of spleen cells. Parameters that affect degree of protection. Diabetes 38:24–30

Carroll SF, Collier RJ (1987) Active site of *Pseudomonas aeruginosa* exotoxin A. Glutamic acid 553 is photolabeled by NAD and shows functional homology with glutamic acid 148 of diphtheria toxin. J Biol Chem 262:8707–8711

Carroll SF, Collier RJ (1988) Amino acid sequence homology between the enzymic domains of diphtheria toxin and *Pseudomonas aeruginosa* exotoxin A. Mol Microbiol 2:293–296

Carroll SF, McCloskey JA, Crain PF, Oppenheimer NJ, Marschner TM, Collier RJ (1985) Photoaffinity labeling of diphtheria toxin fragment A with NAD: structure of the photoproduct at position 148. Proc Natl Acad Sci USA 82:7237–7241

Chang Y-C, Soman G, Graves DJ (1986) Identification of an enzymatic activity that hydrolyzes protein-bound ADP-ribose in skeletal muscle. Biochem Biophys Res Commun 139:932–939

Chao D, Severson DL, Zwiers H, Hollenberg MD (1994) Radiolabelling of bovine myristoylated alanine-rich protein kinase C substrate (MARKS) in an ADP-ribosylation reaction. Biochem Cell Biol 72:391–396

Choe S, Bennett MJ, Fujii G, Curmi PMG, Kantardjieff KA, Collier RJ, Eisenberg D (1992) The crystal structure of diphtheria toxin. Nature 357:216–222

Cieplak W Jr, Mead DJ, Messer RJ, Grant CCR (1995) Site-directed mutagenic alteration of potential active-site residues of the A subunit of *Escherichia coli* heat-labile enterotoxin. J Biol Chem 270:30545–30550

Clancy R, Leszczynska J, Amin A, Levartovsky D, Abramson SB (1995) Nitric oxide stimulates ADP-ribosylation of actin in association with the inhibition of actin polymerization in human neutrophils. J Leukoc Biol 58:196–202

Coburn J, Kane AV, Feig L, Gill DM (1991) *Pseudomonas aeruginosa* exoenzyme S requires a eukaryotic protein for ADP-ribosyltransferase activity. J Biol Chem 266:6438–6446

Coggins PF, McLean K, Nagy A, Zwiers H (1993a) ADP-ribosylation of the neuronal phosphoprotein B-50/GAP-43. J Neurochem 60:368–371

Coggins PJ, McLean K, Zwiers H (1993b) Neurogranin, a B-50/GAP-43-immunoreactive C-kinase substrate (BICKS), is ADP-ribosylated. FEBS Lett 335:109–113

Collier RJ (1990) Diphtheria toxin: structure and function of a cytocidal protein. In: Moss J, Vaughan M (eds) ADP-ribosylating toxins and G proteins: insights into signal transduction. American Society for Microbiology, Washington DC, pp 3–19

Davis T, Shall S (1995) Sequence of a chicken erythroblast mono(ADP-ribosyl)transferase-encoding gene and its upstream region. Gene 164:371–372

De Matteis MA, Di Girolamo M, Colanzi A, Pallas M, De Tullio G, McDonald LJ, Moss J, Santini G, Bannykh S, Corda D, Luini A (1994) Stimulation of endogenous ADP-ribosylation by brefeldin A. Proc Natl Acad Sci USA 91:1114–1118

Di Girolamo M, Silletta MG, De Matteis MA, Braca A, Colanzi A, Pawlak D, Rasenick MM, Luini A, Corda D (1995) Evidence that the 50-kDa substrate of brefeldin A-dependent ADP-ribosylation binds GTP and is modulated by the G-protein $\beta\gamma$ subunit complex. Proc Natl Acad Sci USA 92:7065–7069

Dimmeler S, Lottspeich F, Brune B (1992) Nitric oxide causes ADP-ribosylation and inhibition of glyceraldehyde-3-phosphate dehydrogenase. J Biol Chem 267:16771–16774

Domenighini M, Montecucco C, Ripka WC, Rappuoli R (1991) Computer modelling of the NAD binding site of ADP-ribosylating toxins: active-site structure and mechanism of NAD binding. Mol Microbiol 5:23–31

Domenighini M, Magagnoli C, Pizza M, Rappuoli R (1994) Common features of the NAD-binding and catalytic site of ADP-ribosylating toxins. Mol Microbiol 14:41–50

Donadoni ML, Gavezzotti R, Borella F, De Giulio AM, Gorio A (1995) Experimental diabetic neuropathy. Inhibition of protein mono-ADP-ribosylation prevents reduction of substance P axonal transport. J Pharmacol Exp Ther 274:570–576

Donnelly LE, Boyd RS, MacDermot J (1992) $G_s\alpha$ is a substrate for mono(ADP-ribosyl)transferase of NG108-15 cells. ADP-ribosylation regulates $G_s\alpha$ activity and abundance. Biochem J 288:331–336

Douglas CM, Collier RJ (1987) Exotoxin A of *Pseudomonas aeruginosa*: substitution of glutamic acid-553 with aspartic acid drastically reduces toxicity and enzymic activity. Infect Immun 169:4967–4971

Doukas J, Mordes JP (1993) T lymphocytes capable of activating endothelial cells in vitro are present in rats with autoimmune diabetes. J Immunol 150:1036–1046

Duman RS, Winston SM, Clark JA, Nestler EJ (1990) Corticosterone regulates the expression of ADP-ribosylation factor messenger RNA and protein in rat cerebral cortex. J Neurochem 55:1813–1816

Duman RS, Terwilliger RZ, Nestler EJ (1991) Endogenous ADP-ribosylation in brain: initial characterization of substrate proteins. J Neurochem 57:2124–2132

Ehret-Hilberer S, Nullans G, Aunis D, Virmaux N (1992) Mono-ADP-ribosylation of transducin catalyzed by rod outer segment extract. FEBS Lett 309:394–398

Fangmann J, Schwinzer M, Winkler M, Wonigeit K (1990) Expression of RT6 alloantigens and the T-cell receptor on intestinal intraepithelial lymphocytes of the rat. Transplant Proc 22:2543–2544

Feldman AM, Levine MA, Baughman KL, Van Dop C (1987) NAD$^+$-mediated stimulation of adenylate cyclase in cardiac membranes. Biochem Biophys Res Commun 142:631–637

Finco C, Abbracchio MP, Malosio ML, Cattabeni F, Di Giulio AM, Paternieri B, Mantegazza P, Gorio A (1992) Diabetes-induced alteration of central nervous system G proteins. ADP-ribosylation, immunoreactivity, and gene-expression studies in rat striatum. Mol Chem Neuropathol 17:259–272

Gerber LD, Kodukula K, Udenfriend S (1992) Phosphatidylinositol glycan (PI-G) anchored membrane proteins. Amino acid requirements adjacent to the site of cleavage and PI-G attachment in the COOH-terminal signal peptide. J Biol Chem 267:12168–12173

Godeau F, Belin D, Koide SS (1984) Mono(adenosine diphosphate ribosyl) transferase in *Xenopus* tissues. Direct demonstration by a zymographic localization in sodium dodecyl sulfate-polyacrylamide gels. Anal Biochem 137:287–296

Gorio A, Donadoni ML, Di Giulio AM (1995) Nitric oxide-sensitive protein ADP-ribosylation is altered in rat diabetic neuropathy. J Neurosci Res 40:420–426

Greiner DL, Handler ES, Nakano K, Mordes JP, Rossini AA (1986) Absence of the RT-6 T cell subset in diabetes-prone BB/W rats. J Immunol 136:148–151

Greiner DL, Mordes JP, Handler ES, Angelillo M, Nakamura N, Rossini AA (1987) Depletion of RT6.1$^+$ T lymphocytes induces diabetes in resistant Biobreeding/Worcestter (BB/W) rats. J Exp Med 166:461–475

Grimaldi JC, Balasubramanian S, Kabra NH, Shanafelt A, Bazan JF, Zurawski G, Howard MC (1995) CD38-mediated ribosylation of proteins. J Immunol 155:811–817

Haag F, Koch F, Thiele H-G (1990) Polymorphism between not T-cell alloantigens RT6.1 and RT6.2 is based on multiple amino acid substitutions. Transplant Proc 22:2541–2542

Haag F, Nolte F, Hollmann C, Thiele H-C (1993) Analysis of the gene for the rat T-cell alloantigen RT6: evidence for alternative splicing in the 5′ region. Transplant Proc 25:2784–2785

Haag F, Koch-Nolte F, Kuhl M, Lornezen S, Thiele H-G (1994) Premature stop codons inactivate the RT6 genes of the human and chimpanzee species. J Mol Biol 243:537–546

Haag F, Andresen V, Karsten S, Koch-Nolte F, Thiele H-G (1995) Both allelic forms of the rat T cell differentiation marker RT6 display nicotinamide adenine dinucleotide (NAD)-glycohydrolase activity, yet only RT6.2 is capable of automodification upon incubation with NAD. Eur J Immunol 25:2355–2361

Han XY, Galloway DR (1995) Active site mutations of *Pseudomonas aeruginosa* exotoxin A. Analysis of the His^{440} residue. J Biol Chem 270:679–684

Hawkins DJ, Browning ET (1982) Tubulin adenosine diphosphate ribosylation is catalyzed by cholera toxin. Biochem 21:4474–4479

Howard MC, Grimaldi JC Bazan JF, Lund FE, Santos-Argumedo L, Parkhouse RME, Walseth TF, Lee HC (1993) Formation and hydrolysis of cyclic ADP-ribose catalyzed by lymphocyte antigen CD38. Science 262:1056–1059

Jacobson MK, Loflin PT, Aboul-Ela N, Mingmuang M, Moss J, Jacobson EL (1990) Modification of plasma membrane protein cysteine residues by ADP-ribose in vivo. J Biol Chem 265:10825–10828

Johnson VG, Nicholls P (1994) Histidine-21 does not play a major role in diphtheria toxin catalysis. J Biol Chem 269:4349–4354

Jung M, Just I, van Damme J, Vandekerckhove J, Aktories K (1993) NAD-binding site of the C3-like ADP-ribosyltransferase from Clostridium limosum. J Biol Chem 268:23215–23218

Just I, Wollenberg P, Moss J, Aktories K (1994) Cysteine-specific ADP-ribosylation of actin. Eur J Biochem 221:1047–1054

Just I, Sehr P, Jung M, van Damme J, Puype M, Vandekerckhove J, Moss J, Aktories K (1995a) ADP-ribosyltransferase type A from turkey erythrocytes modifies actin at arg-95 and arg-372. Biochemistry 34:326–333

Just I, Selzer J, Jung M, van Damme J, Vandekerckhove J, Aktories K (1995b) Rho-ADP-ribosylating exoenzyme from Bacillus cereus. Purification, characterization, and identification of the NAD-binding site. Biochem 34:334–340

Kaslow HR, Schlotterbeck JD, Mar VL, Burnette NW (1989) Alkylation of cysteine 41, but not cysteine 200, decreases the ADP-ribosyltransferase activity of the S1 subunit of pertussis toxin. J Biol Chem 264:6386–6390

Kessler SP, Galloway DR (1992) Pseudomonas aeruginosa Exotoxin A interaction with eucaryotic elongation factor 2. Role of the His[426] residue. J Biol Chem 267:19107–19111

Kharadia SV, Huiatt TW, Huang H-Y, Peterson JE, Graves DJ (1992) Effect of an arginine-specific ADP-ribosyltransferase inhibitor on differentiation of embryonic chick skeletal muscle cells in culture. Exp Cell Res 201:33–42

Klausner RD, Donaldson JG, Lippincott-Schwartz J (1992) Brefeldin A: insights into the control of membrane traffic and organelle structure. J Cell Biol 116:1071–1080

Klebl BM, Matsushita S, Pette D (1994) Localization of an arginine-specific mono-ADP-ribosyltransferase in skeletal muscle sarcolemma and transverse tubules. FEBS Lett 342:66–70

Knight DA, Finck-Barbancon V, Kulich SM, Barbieri JT (1995) Functional domains of Pseudomonas aeruginosa exoenzyme S. Infect Immunol 63:3182–3186

Koch F, Kashan A, Thiele H-G (1988) The rat T-cell differentiation marker RT6.1 is more polymorphic than its alloantigenic counterpart RT6.2. Immunology 65:259–265

Koch F, Haag F, Kashan A, Thiele H-G (1990a) Primary structure of rat RT6.2, a nonglycosylated phosphatidylinositol-linked surface marker of postthymic T cells. Proc Natl Acad Sci USA 87:964–967

Koch F, Haag F, Thiele H-G (1990b) Nucleotide and deduced amino acid sequence for the mouse homologue of the rat T-cell differentiation marker RT6. Nucleic Acids Res 18:3636

Koch-Nolte F, Haag F, Kuhl M, van Heyningen V, Hoovers J, Grzeschik K-H, Singh S, Thiele H-G (1993) Assignment of the human RT6 gene to 11q13 by PCR screening of somatic cell hybrids and in situ hybridization. Genomics 18:404–406

Koch-Nolte F, Hollmann C, Kuhl M, Haag F, Prochazka M, Leiter E, Thiele H-G (1995a) Molecular polymorphism in the Rt6 genes of laboratory mice correlates with the allotypes of the H1 minor histocompatability system. Immunogenetics 41:152–155

Koch-Nolte F, Klein J, Hollmann C, Kuhl M, Haag F, Gaskins HR, Leiter E, Thiele H-G (1995b) Defects in the structure and expression of the genes for the T cell marker RT6 in NZW and (NZW X NZW)F₁ mice. Internatl Immunol 7:883–890

Koch-Nolte F, Petersen D, Balasubramanian S, Haag F, Kahlke D, Willer T, Kastelein R, Bazan F, Thiele H-G (1996) Mouse T cell membrane proteins Rt6-1 and Rt6-2 are arginine/protein mono (ADP-ribosyl) transferases and share secondary structure motifs with ADP-ribosylating bacterial toxins. J Biol Chem 271:7686–7693

Koch T, Ruger W (1994) The ADP-ribosyltransferases (gpAlt) of bacteriophages T2, T4, and T6: sequencing of the genes and comparison of their products. Virology 203:294–298

Kots AY, Skurat AV, Sergienko EA, Bulargina TV, Severin ES (1992) Nitroprusside stimulates the cysteine specific mono(ADP-ribosylation) of glyceraldehyde-3-phosphate dehydrogenase from human erythrocytes. FEBS Lett 300:9–12

Kulich SM, Yahr TL, Mende-Mueller LM, Barbieri JT, Frank DW (1994) Cloning the structural gene for the 49-kDa form of exoenzyme S (exoS) from Pseudomonas aeruginosa strain 388. J Biol Chem 269:10431–10437

Lee HC, Walseth TF, Bratt GT, Hayes RN, Clapper DL (1989) Structural determination of a cyclic metabolite of NAD⁺ with intracellular Ca²⁺-mobilizing activity. J Biol Chem 264:1608–1611

Liu Y, Kahn ML (1995) ADP-ribosylation of Rhizobium meliloti glutamine synthetase III in vivo. J Biol Chem 270:1624–1628

Lobet Y, Cluff CW, Cieplak W Jr (1991) Effect of site-directed mutagenic alterations on ADP-ribosyltransferase activity of the A subunit of Escherichia coli heat-labile enterotoxin. Infect Immun 59:2870–2879

Ludden PW (1994) Reversible ADP-ribosylation as a mechanism of enzyme regulation in procaryotes. Mol Cell Biochem 138:123–129

Maehama T, Nishina H, Katada T (1994) ADP-ribosylarginine glycohydrolase catalyzing the release of ADP-ribose from the cholera toxin-modified α-subunits of GTP-binding proteins. J Biochem 116:1134–1138

Maehama T, Nishina H, Hoshino S, Kanaho Y, Katada T (1995) NAD⁺-dependent ADP-ribosylation of T lymphocyte alloantigen RT6.1 reversibly proceeding in intact rat lymphocytes. J Biol Chem 270:22747–22751

Marsischky GT, Ikejima M, Suzuki H, Sugimura T, Esumi H, Miwa M, Collier RJ (1992) Directed mutagenesis of glutamic acid 988 of poly(ADP-ribose) polymerase. In:

Poirier GG, Moreau P (eds) ADP-ribosylation reactions. Springer, Berlin Heidelberg New York, pp 47–52

Matsuura R, Tanigawa Y, Tsuchiya M, Mishima K, Yoshimura Y, Shimoyama M (1988) ADP-ribosylation suppresses phosphorylation of the L-type pyruvate kinase. Biochem Biophys Acta 969:57–65

Matsuyama S, Tsuyama S (1991) Mono-ADP-ribosylation in brain: purification and characterization of ADP-ribosyltransferases affecting actin from rat brain. J Neurochem 57:1380–1387

McDonald LJ, Moss J (1993a) Nitric oxide-independent, thiol-associated ADP-ribosylation inactivates aldehyde dehydrogenase. J Biol Chem 268:17878–17882

McDonald LJ, Moss J (1993b) Stimulation by nitric oxide of a novel linkage of NAD to glyceraldehyde 3-phosphate dehydrogenase. Proc Natl Acad Sci USA 90:6238–6241

McDonald LJ, Moss J (1994) Enzymatic and nonenzymatic ADP-ribosylation of cysteine. Mol Cell Biochem 138:221–226

McDonald LJ, Wainschel LA, Oppenheimer NJ, Moss J (1992) Amino acid-specific ADP-ribosylation: structural characterization and chemical differentiation of ADP-ribose-cysteine adducts formed nonenzymatically and in a pertussis toxin-catalyzed reaction. Biochem 31:11881–11887

McMahon KK, Piron KJ, Ha VT, Fullerton AT (1993) Developmental and biochemical characteristics of the cardiac membrane-bound arginine-specific mono-ADP-ribosyltransferase. Biochem J 293:789–793

Merritt EA, Sarfaty S, Pizza M, Domenighini M, Rappuoli R, Hol WGJ (1995) Mutation of a buried residue causes loss of activity but no conformational change in the heat-labile enterotoxin of *Escherichia coli*. Struct Biol 2:269–272

Mishima K, Tsuchiya M, Tanigawa Y, Yoshimura Y, Shimoyama M (1989) DNA-dependent mono(ADP-ribosyl)ation of p33, an acceptor protein in hen liver nuclei. Eur J Biochem 179:267–273

Mishima K, Terashima M, Obara S, Yamada K, Imai K, Shimoyama M (1991) Arginine-specific ADP-ribosyltransferase and its acceptor protein p33 in chicken polymorphonuclear cells: co-localization in the cell granules, partial characterization, and *in situ* mono(ADP-ribosyl)ation. J Biochem 110:388–394

Mojcik CF, Greiner DL, Medlock ES, Komschlies KL, Goldschneider I (1988) Characterization of RT6 bearing rat lymphocytes. I. Ontogeny of the RT6[+] subset. Cell Immunol 114:336–346

Molina y Vedia L, Nolan RD, Lapetina EG (1989) The effect of iloprost on the ADP-ribosylation of $G_s\alpha$ (the α-subunit of G_s). Biochem J 261:841–845

Moss J, Stanley SJ (1981a) Histone-dependent and histone-independent forms of an ADP-ribosyltransferase from human and turkey erythrocytes. Proc Natl Acad Sci USA 78:4809–4812

Moss J, Stanley SJ (1981b) Amino acid-specific ADP-ribosylation. Identification of an arginine-dependent ADP-ribosyltransferase in rat liver. J Biol Chem 256:7830–7833

Moss J, Vaughan M (1978) Isolation of an avian erythrocyte protein possessing ADP-ribosyltransferase activity and capable of activating adenylate cyclase. Proc Natl Acad Sci USA 75:3621–3624

Moss J, Vaughan M (1988) ADP-ribosylation of guanyl nucleotide-binding proteins by bacterial toxins. Adv Enzymol 61:303–379

Moss J, Vaughan M (eds) (1990) ADP-ribosylating toxins and G proteins: insights into signal transduction. American Society for Microbiology, Washington DC

Moss J, Vaughan M (1995) Structure and function of ARF proteins: activators of cholera toxin and critical components of intracellular vesicular transport process. J Biol Chem 270:12327–12330

Moss J, Stanley SJ, Oppenheimer NJ (1979) Substrate specificity and partial purification of a stereospecific NAD-and guanidine-dependent ADP-ribosyltransferase from avian erythrocytes. J Biol Chem 254:8891–8894

Moss J, Stanley SJ, Watkins PA (1980) Isolation and properties of an NAD- and guanidine-dependent ADP-ribosyltransferase from turkey erythrocytes. J Biol Chem 255:5838–5840

Moss J, Stanley SJ, Osborne JC Jr (1981) Effect of self-association on activity of an ADP-ribosyltransferase from turkey erythrocytes. J Biol Chem 256:11452–11456

Moss J, Stanley SJ, Osborne JC Jr (1982) Activation of NAD:arginine ADP-ribosyltransferase by histone. J Biol Chem 257:1660–1663

Moss J, Osborne JC Jr, Stanley SJ (1984a) Activation of an erythrocyte NAD:arginine ADP-ribosyltransferase by lysolecithin and nonionic and zwitterionic detergents. Biochemistry 23:1353–1357

Moss J, Watkins PA, Stanley SJ, Purnell MR, Kidwell WR (1984b) Inactivation of glutamine synthetases by an NAD:arginine ADP-ribosyltransferase. J Biol Chem 259:5100–5104

Moss J, Jacobson MK, Stanley SJ (1985) Reversibility of arginine-specific mono(ADP-ribosyl)ation: identification in erythrocytes of an ADP-ribose-L-arginine cleavage enzyme. Proc Natl Acad Sci USA 82:5603–5607

Moss J, Oppenheimer NJ, West RE Jr, Stanley SJ (1986) Amino acid specific ADP-ribosylation: substrate specificity of an ADP-ribosylarginine hydrolase from turkey erythrocytes. Biochemistry 25:5408–5414

Moss J, Tsai S-C, Adamik R, Chen H-C, Stanley SJ (1988) Purification and characterization of ADP-ribosylarginine hydrolase from turkey erythrocytes. Biochemistry 27:5819–5823

Moss J, Stanley SJ, Levine RL (1990) Inactivation of bacterial glutamine synthetase by ADP-ribosylation. J Biol Chem 265:21056–21060

Moss J, Stanley SJ, Nightingale MS, Murtagh JJ Jr, Monaco L, Mishima K, Chen H-C, Williamson KC, Tsai S-C (1992) Molecular and Immunological characterization of ADP-ribosylarginine hydrolases. J Biol Chem 267:10481–10488

Moss J, Stanley SJ, Vaughan M, Tsuji T (1993) Interaction of ADP-ribosylation factor with Escherichia coli enterotoxin that contains an inactivation lysine 112 substitution. J Biol Chem 268:6383–6387

Narumiya S, Sekine A, Fujiwara M (1988) Substrate for botulinum ADP-ribosyltransferase, Gb, has an amino acid sequence homologous to a putative *rho* gene product. J Biol Chem 263:17255–17257

Nemoto Y, Namba T, Kozaki S, Narumiya S (1991) *Clostridium botulinum* C3 ADP-ribosyltransferase gene. Cloning sequencing, and expression of a functional protein in *Escherichia coli*. J Biol Chem 266:19312–19319

Ness SA, Marknell A, Graf T (1989) The *v-myb* oncogene product binds to and activates the promyelocyte-specific *mim-1* gene. Cell 59:1115–1125

Nestler EJ, Terwilliger RZ, Duman RS (1995) Regulation of endogenous ADP-ribosylation by acute and chronic lithium in rat brain. J Neurochem 64:2319–2324

Obara S, Yamada K, Yoshimura Y, Shimoyama M (1991) Evidence for the endogenous GTP-dependent ADP-ribosylation of the α-subunit of the stimulatory guanyl-nucleotide-binding protein concomitant with an increase in basal adenylyl cyclase activity in chicken spleen cell membrane. Eur J Biochem 200:75–80

Okazaki IJ, Zolkiewska A, Nightingale MS, Moss J (1994) Immunological and structural conservation of mammalian skeletal muscle glycosylphosphatidylinositol-linked ADP-ribosyltransferases. Biochemistry 33:12828–12836

Okazaki IJ, Kim H-J, McElvaney G, Lesma E, Moss J (1996a) Molecular characterization of a glycosylphosphatidylinositol-linked ADP-ribosyltransferase from lymphocytes. Blood, in press

Okazaki IJ, Kim H-J, Moss J (1996b) A novel membrane-bound lymphoxyte ADP-ribosyltransferase cloned from Yac-1 cells. J Biol Chem, in press

Oppenheimer NJ (1978) Structural determination and stereospecificity of the choleragen-catalyzed reaction of NAD^+ with guanidines. J Biol Chem 253:4907–4910

Osborne JC Jr, Stanley SJ, Moss J (1985) Kinetic mechanisms of two NAD:arginine ADP-ribosyltransferases: the soluble, salt-stimulated transferase from turkey erythrocytes and choleragen, a toxin from *Vibrio cholera*. Biochemistry 24:5235–5240

Papini E, Schiavo G, Sandona D, Rappuoli R, Montecucco C (1989) Histidine 21 is at the NAD^+ binding site of diphtheria toxin. J Biol Chem 264:12385–12388

Papini E, Santucci A, Schiavo G, Domenighini M, Neri P, Rappuoli R, Montecucco R (1991) Tyr-65 is photolabelled by 8-azido adenine and 8-azido-adenosine at the NAD binding site of diphtheria toxin. J Biol Chem 266:2494–2498

Pelham HRB (1991) Multiple targets for brefeldin A. Cell 67:449–451

Peterson JE, Larew JS-A Graves DJ (1990) Purification and partial characterization of arginine-specific ADP-ribosyltransferase from skeletal muscle microsomal membranes. J Bio Chem 265:17062–17069

Piron KJ, McMahon KK (1990) Localization and partial characterization of ADP-ribosylation products in hearts from adult and neonatal rats. Biochem J 270:591–597

Pizza M, Bartoloni A, Prugnola A, Silvestri S, Rappuoli R (1988) Subunit S1 of pertussis toxin: mapping of the regions essential for ADP-ribosyltransferase activity. Proc Natl Acad Sci USA 85:7521–7525

Pizza M, Domenighini M, Hol W, Giannelli V, Fontana MR, Giuliani MM, Magagnoli C, Peppoloni S, Manetti R, Rappuoli R (1994) Probing the structure-activity relationship of *Escherichia coli* LT-A by site-directed mutagenesis. Mol Microbiol 14:51–60

Pozdnyakov N, Lloyd A, Reddy VN, Sitaramayya A (1993) Nitric oxide-regulated endogenous ADP-ribosylation of rod outer segment proteins. Biochem Biophys Res Commun 192:610–615

Prochazka M, Leiter EH, Serreze DV, Coleman DL (1987) Three recessive loci required for insulin-dependent diabetes in nonobese diabetic mice. Science 237:286–289

Prochazka M, Gaskins HR, Leiter EH, Koch-Nolte F, Haag F, Thiele H-G (1991) Chromosomal localization, DNA polymorphism, and expression of *Rt-6*, the mouse homologue of rat T-lymphocyte differentiation marker *RT6*. Immunogenetics 33:152–156

Quist EE, Coyle DL, Vasan R, Satumitra N, Jacobson EL, Jacobson MK (1994) Modification of cardiac membrane adenylate cyclase activity and $G_{s\alpha}$ by NAD and endogenous ADP-ribosyltransferase. J Mol Cell Cardiol 26:251–160

Raffaelli N, Scaife RM, Purich DL (1992) ADP-ribosylation of chicken red cell tubulin and inhibition of microtubule self-assembly in vitro by the NAD^+-dependent avian ADP-ribosyltransferase. Biochem Biophys Res Commun 184:414–418

Rankin PW, Jacobson EL, Benjamin RC, Moss J, Jacobson MK (1989) Quantitative studies of inhibitors of ADP-ribosylation *in vitro* and *in vivo*. J Biol Chem 264:4312–4317

Rappuoli R, Pizza M (1991) Structure and evolutionary aspects of ADP-ribosylating toxins. In: Alouf JE, Freer JH (eds) Sourcebook of bacterial protein toxins. Academic, San Diego, pp 1–21

Rigby M, Bortell R, Stevens LA, Moss J, Kanaitsuka T, Shigeta H, Mordes JP, Greiner DL, Rossini AA (1996) Rat RT6.2 and mouse Rt6 locus 1 are NAD:arginine ADP-ribosyltransferases with auto-ADP-ribosylation activity. J Immunol, 156:4259–4265

Rosa JL, Perez JX, Ventura F, Tauler A, Gil J, Shimoyama M (1995) Role of the N-terminal region in covalent modification of 6-phosphofructo-2-kinase/fructose-2,6-bisphosphatase: comparison of phosphorylation and ADP-ribosylation. Biochem J 309:119–125

Rossini AA, Mordes JP, Greiner DL, Nakano K, Appel MC, Handler ES (1986) Spleen cell transfusion in the Bio-Breeding/Worcester rat. Prevention of diabetes, major histocompatability complex restriction, and long-term persistence of transfused cells. J Clin Invest 77:1399–1401

Saito N, Guitart X, Hayward MD, Tallman JF, Duman RS, Nestler EJ (1989) Corticosterone differentially regulates the expression of $G_{s\alpha}$ and $G_{i\alpha}$ messenger RNA and protein in rat cerebral cortex. Proc Natl Acad Sci USA 86:3906–3910

Scaife RM, Wilson L, Purich DL (1992) Microtubule protein ADP-ribosylation in vitro leads to assembly inhibition and rapid depolymerization. Biochemistry 31:310–316

Schering B, Barmann M, Chhatwal GS, Geipel U, Aktories K (1988) ADP-ribosylation of skeletal muscle and non-muscle actin by *Clostridium perfringens* iota toxin. Eur J Biochem 171:225–229

Schuman EM, Meffert MK, Schulman H, Madison DV (1994) An ADP-ribosyltransferase as a potential target for nitric oxide action in hippocampal long-term potentiation. Proc Natl Acad Sci USA 91:11958–11962

Sekine A, Fujiwara M, Narumiya S (1989) Asparagine residue in the *rho* gene product is the modification site for botulinum ADP-ribosyltransferase. J Biol Chem 264:8602–8605

Sheffler LA, Wink DA, Melilo G, Cox GW (1995) Characterization of nitric oxide-stimulated ADP-ribosylation of various proteins from the mouse macrophage cell line ANA-1 using sodium nitroprusside and the novel nitric oxide-donating compound diethlamine dinitric oxide. J Leukoc Biol 57:152–159

Silman NJ, Carr NG, Mann NH (1995) ADP-ribosylation of glutamine synthetase in the cyanobacterium *Synechocystis* sp. strain PCC 6803. J Bacteriol 177:3527–3533

Simonin F, Poch O, Delarue M, de Murcia G (1993) Identification of potential active-site residues in the human poly(ADP-ribose) polymerase. J Biol Chem 268:8529–8535

Sixma TK, Pronk SE, Kalk KH, Wartna ES, van Zanten BAM, Witholt B, Hol WGJ (1991) Crystal structure of a cholera toxin-related heat-labile enterotoxin from *E. coli*. Nature 351:371–377

Sixma TK, Kalk KH, van Zanten BAM, Dauter Z, Kingma J, Witholt B, Hol WGJ (1993) Refined structure of *Escherichia coli* heat-labile enterotoxin, a close relative of cholera toxin. J Mol Biol 230:890–918

Smith KP, Benjamin RC, Moss J, Jacobson MK (1985) Identification of enzymatic activities which process protein bound mono(ADP-ribose). Biochem Biophys Res Commun 126:136–142

Soman G, Mickelson JR, Louis CF, Graves DJ (1984) NAD:guanidino group-specific mono-ADP-ribosyltransferase activity in skeletal muscle. Biochem Biophys Res Commun 120:973–980

Soman G, Haregewoin A, Hom RC, Finberg RW (1991) Guanidine group specific ADP-ribosyltransferase in murine cells. Biochem Biophys Res Commun 176:301–308

Song WK, Wang W, Foster RF, Bielser DA, Kaufamn SJ (1992) H-36-alpha 7 is a novel integrin alpha chain that is developmentally regulated during skeletal myogenesis. J Cell Biol 117:643–657

Stein PE, Boodhoo A, Armstrong GD, Cockle SA, Klein MH, Read RJ (1994) The crystal structure of pertussis toxin. Structure 2:45–57

Sternweis PC, Robishaw JD (1984) Isolation of two proteins with high affinity for guanine nucleotides from membrane of bovine brain. J Biol Chem 259:13806–13813

Takada T, Iida K, Moss J (1993) Cloning and site-directed mutagenesis of human ADP-ribosylarginine hydrolase. J Biol Chem 268:17837–17843

Takada T, Iida K, Moss J (1994) Expression of NAD glycohydrolase activity by rat mammary adenocarcinoma cells transformed with rat T cell alloantigen RT6.2. J Biol Chem 269:9420–9423

Takada T, Iida K, Moss J (1995) Conservation of a common motif in enzymes catalyzing ADP-ribose transfer. J Biol Chem 270:541–544

Takenaka S, Nakano Y, Tsuyama S (1994) Mono-ADP-ribosylation of microtubule-associated protein 2 that inhibits polymerization of rat brain microtubules. In: The 11th international symposium on ADP-ribosylation. DNA repair, signal transduction. Abstract no 56. Strasbourg-Bischenberg, France

Tamir A, Gill D (1988) ADP-ribosylation by cholera toxin of membranes derived from brain modifies the interaction of adenylate cyclase with guanine nucleotides and NaF. J Neurochem 50:1791–1797

Tanigawa Y, Tsuchiya M, Imai Y, Shimoyama M (1983a) Mono(ADP-ribosyl)ation of hen liver nuclear proteins suppresses phosphorylation. Biochem Biophys Res Commun 113:135–141

Tanigawa Y, Tsuchiya M, Imai Y, Shimoyama M (1983b) ADP-ribosylation regulates the phosphorylation of histones by the catalytic subunit of cyclic AMP-dependent protein kinase. FEBS Lett 160:217–220

Tanigawa Y, Tsuchiya M, Imai Y, Shimoyama M (1984) ADP-ribosyltransferase from hen liver nuclei. J Biol Chem 259:2022–2029

Tanuma S, Endo H (1990) Identification in human erythrocytes of mono(ADP-ribosyl) protein hydrolase that cleaves a mono(ADP-ribosyl) G_i linkage. FEBS Lett 261:381–384

Tanuma S, Kawashima K, Endo H (1987) An NAD:cysteine ADP-ribosyltransferase is present in human erythrocytes. J Biochem 101:821–824

Tanuma S, Kawashima K, Endo H (1988) Eukaryotic mono(ADP-ribosyl)transferase that ADP-ribosylates GTP-binding regulatory G_i protein. J Biol Chem 263:5485–5489

Terashima M, Mishima K, Yamada K, Tsuchiya M, Wakutani T, Shimoyama M (1992) ADP-ribosylation of actins by arginine-specific ADP-ribosyltransferase purified from chicken heterophils. Eur J Biochem 204:305–311

[Terashima M, Yamamori C, Shimoyama M (1995) ADP-ribosylation of Arg28 and Arg206 on the actin molecule by chicken arginine-specific ADP-ribosyltransferase. Eur J Biochem 231:242–249]

Thiele H-G, Koch F, Hamann A, Arndt R (1986) Biochemical characterization of the T-cell alloantigen RT6.2. Immunology 59:195–201

Thiele H-G, Koch F, Kashan A (1987) Postnatal distribution profiles of Thy-1[+] and RT6[+] cells in peripheral lymph nodes of DA-rats. Transplant Proc 19:3157–3160

Thiele H-G, Haag F, Nolte F (1993) Asymmetric expression of RT6.1 and RT6.2 alloantigens in (RT6[a] X RT6[b])F_1 rats is due to a pretranslational mechanism. Transplant Proc 25:2786–2788

Tsai S-C, Adamik R, Moss J, Vaughan M, Manne V, Kung H-F (1985) Effects of phospholipids and ADP-ribosylation on GTP hydrolysis by *Escherichia coli*-synthesized Ha-*ras*-encoded p21. Proc Natl Acad Sci USA 82:8310–8314

Tsai S-C, Adamik R, Haun RS, Moss J, Vaughan M (1993) Effects of brefeldin A and accessory proteins on association of ADP-ribosylation factors 1, 3, and 5 with Golgi. J Biol Chem 268:10820–10825

Tsuchiya M, Tanigawa Y, Ushiroyama T, Matsuura R, Shimoyama M (1985) ADP-ribosylation of phosphorylase kinase and block of phosphate incorporation into the enzyme. Eur J Biochem 147:33–40

Tsuchiya M, Hara N, Yamada K, Osago H, Shimoyama M (1994) Cloning and expression of cDNA for arginine-specific ADP-ribosyltransferase from chicken bone marrow cells. J Biol Chem 269:27451–27457

Tsuchiya M, Osago H, Shimoyama M (1995) A newly identified GPI-anchored arginine-specific ADP-ribosyltransferase activity in chicken spleen. Biochem Biophys Res Commun 214:760–764

Tsuji T, Inoue T, Miyama A, Okamoto K, Honda T, Miwatani T (1990) A single amino acid substitution in the A subunit of *Escherichia coli* enterotoxin results in a loss of its toxic activity. J Biol Chem 265:22520–22525

Tweten RK, Barbieri JT, Collier RJ (1985) Diphtheria toxin. Effect of substituting aspartic acid for glutamic acid 148 on ADP-ribosyltransferase activity. J Biol Chem 260:10392–10394

Uchikoshi F, Ito T, Kamiike W, Moriguchi A, Nozaki S, Ito A, Kuhara A, Miyata M, Matsuda H, Miyasaka M, Nakao H, Makino S, Nozawa M (1995) Appearance of immunoregulatory RT6[+] T cells after successful pancreas transplantation in diabetic BB rats. Transplant Proc 27:599–601

Ui M (1990) Pertussis toxin as a valuable probe for G-protein involvement in signal transduction. In: Moss J, Vaughan M (eds) ADP-ribosylating toxins and G proteins: insights into signal transduction. American Society for Microbiology, Washington DC, pp 45–77

Uroshiyama T, Tanigawa Y, Tsuchiya M, Matsuura R, Ueki M, Sugimoto O, Shimoyama M (1985) Amino acid sequence of histone H1 at the ADP-ribose-accepting site and ADP-ribose·histone-H1 adduct as an inhibitor of cyclic-AMP-dependent phosphorylation. Eur J Biochem 151:173–177

Vandekerckhove J, Schering B, Barmann M, Aktories K (1987) *Clostridium perfringens* iota toxin ADP-ribosylates skeletal muscle actin in Arg-177. FEBS Lett 225:48–52

Vandekerckhove J, Schering B, Barmann M, Aktories K (1988) Botulinum C2 toxin ADP-ribosylates cytoplasmic βγ-actin in arginine 177. J Biol Chem 263:696–700

Wang J, Nemoto E, Kots AY, Kaslow HR, Dennert G (1994) Regulation of cytotoxic T cells by ecto-nicotinamide adenine dinucleotide (NAD) correlates with cell surface GPI-anchored/arginine ADP-ribosyltransferase. J Immunol 153:4048–4058

Wang J, Nemoto E, Dennert G (1996) Regulation of CTL by ectonicotinamide adenine dinucleotide (NAD) involves ADP-ribosylation of a p56[lck]-associated protein. J Immunol 156:2819–2827

Watkins PA, Moss J (1982) Effects of nucleotides on activity of a purified ADP-ribosyltransferase from turkey erythrocytes. Arch Biochem Biophys 216:74–80

Watkins PA, Kanoho Y Moss J (1987) Inhibition of the GTP-ase activity of transducin by an NAD[+]: arginine ADP-ribosyltransferase from turkey erythrocytes. Biochem J 248:749–754

Wegner A, Aktories K (1988) ADP-ribosylated actin caps the barbed ends of actin filaments. J Biol Chem 263:13739–13742

Welsh CF, Moss J, Vaughan M (1994) ADP-ribosylation factors: a family of ~ 20-kDa guanine nucleotide-binding proteins that activate cholera toxin. Mol Cell Biochem 138:157–166

West RE Jr, Moss J (1986) Amino acid specific ADP-ribosylation: specific NAD:arginine mono-ADP-ribosyltransferases associated with turkey erythrocyte nuclei and plasma membranes. Biochemistry 25:8057–8062

Wick MJ, Iglewski BH (1990) *Pseudomonas aeruginosa* exotoxin A. In: Moss J, Vaughan M (eds) ADP-ribosylating toxins and G proteins: insights into signal transduction. American Society for Microbiology, Washington DC, pp 31–43

Williamson KC, Moss J (1990) Mono-ADP-ribosyltransferases and ADP-ribosylarginine hydrolases: a mono-ADP-ribosylation cycle in animal cells. In: Moss J, Vaughan M (eds) ADP-ribosylating toxins and G proteins: insights into signal transduction. American Society for Microbiology, Washington DC, pp 493–510

Wilson BA, Blanke SR, Reich KA, Collier RJ (1994) Active-site mutations of diphtheria toxin. J Biol Chem 269:23296–23301

Wozniak DJ, Hsu L-H, Galloway DR (1988) His-426 of the *Pseudomonas aeruginosa* exotoxin A is required for ADP-ribosylation of elongation factor II. Proc Natl Acad Sci USA 85:8880–8884

Xu Y, Barbancon-Finck V, Barbieri JT (1994) Role of histidine 35 of the S1 subunit of pertussis toxin in the ADP-ribosylation of Transducin. J Biol Chem 269:9993–9999

Yamada K, Tsuchiya M, Mishima K, Shimoyama M (1992) p33, and endogenous target protein for arginine-specific ADP-ribosyltransferase in chicken polymorphonuclear leukocytes, is highly homologous to *mim-1* protein (*myb*-induced myeloid protein-1). FEBS Lett 311:203–205

Yamada K, Tsuchiya M, Nishikori Y, Shimoyama M (1994) Automodification of ar-ginine-specific ADP-ribosyltransferase purified from chicken peripheral heterophils and alteration of the transferase activity. Arch Biochem Biophys 308:31–36

[Yost DA, Moss J (1983) Amino acid-specific ADP-ribosylation. Evidence for two distinct NAD:arginine ADP-ribosyltransferases in turkey erythrocytes. J Biol Chem 258:4926–4929]

Zoche M, Koch K-W (1995) Purified retinal nitric oxide synthase enhances ADP-ri-bosylation of rod outer segment proteins. FEBS Lett 357:178–182

Zolkiewska A, Moss J (1993) Integrin α7 as substrate for a glycosylphosphatidylino-sitol-anchored ADP-ribosyltransferase on the surface of skeletal muscle cells. J Biol Chem 268:25273–25276

Zolkiewska A, Moss J (1995) Processing of ADP-ribosylated integrin 7 in skeletal muscle myotubes. J Biol Chem 270:9227–9233

Zolkiewska A, Nightingale MS, Moss J (1992) Molecular characterization of NAD:ar-ginine ADP-ribosyltransferase from rabbit skeletal muscle. Proc Natl Acad Sci USA 89:11352–11356

Springer-Verlag
and the Environment

We at Springer-Verlag firmly believe that an international science publisher has a special obligation to the environment, and our corporate policies consistently reflect this conviction.

We also expect our business partners – paper mills, printers, packaging manufacturers, etc. – to commit themselves to using environmentally friendly materials and production processes.

The paper in this book is made from low- or no-chlorine pulp and is acid free, in conformance with international standards for paper permanency.